Encyclopedia of Insecticides: Advanced Methodologies

Volume III

Encyclopedia of Insecticides: Advanced Methodologies
Volume III

Edited by **Nancy Cahoy**

New York

Published by Callisto Reference,
106 Park Avenue, Suite 200,
New York, NY 10016, USA
www.callistoreference.com

Encyclopedia of Insecticides: Advanced Methodologies
Volume III
Edited by Nancy Cahoy

International Standard Book Number: 978-1-63239-264-0 (Hardback)

Contents

Preface

The purpose of the book is to provide a glimpse into the dynamics and to present opinions and studies of some of the scientists engaged in the development of new ideas in the field from very different standpoints. This book will prove useful to students and researchers owing to its high content quality.

The aim of this book is to educate researchers, scientists, students and end users (farmers, hobby producers) about insecticides and their usage. The book examines topics related to human health and insecticides, and the relation of insecticides to environments. A special focus on insecticides against pests of urban areas, forests and farm animals, and other developments in pest control in the recent past is also examined.

At the end, I would like to appreciate all the efforts made by the authors in completing their chapters professionally. I express my deepest gratitude to all of them for contributing to this book by sharing their valuable works. A special thanks to my family and friends for their constant support in this journey.

Editor

Insecticides and Human Health

DDT as Anti-Malaria Tool:
The Bull in the China Shop
or the Elephant in the Room?

Mauro Prato, Manuela Polimeni and
Giuliana Giribaldi

Additional information is available at the end of the chapter

1. Introduction

Malaria is a parasitic disease confined mostly to the tropical areas, caused by *Plasmodium* parasites and transmitted by *Anopheles* mosquitoes. In 2010, nearly 655.000 human deaths, mainly of children ≤5 years of age, were registered among more than 200 million cases worldwide of clinical malaria; the vast majority of cases occurred in the African Region (81%) and South-East Asia (13%), and 91% of them were due to *P. falciparum*, the most virulent among *Plasmodia* strains (WHO, 2011a).

In order to achieve malaria eradication, an ambitious objective which has been prosecuted since 2007 by the Bill and Melinda Gates Foundation, the World Health Organization (WHO) and the Roll Back Malaria association, several strategies are currently adopted, and a major role is played by vector control (Roberts & Enserink, 2007; Greenwood, 2008; Khadjavi et al., 2010; Prato et al., 2012). Dichlorodiphenyltrichloroethane (DDT), one of the insecticides recommended by the WHO for indoor residual spraying or treated bednets approaches against *Anopheles* mosquitoes, is currently used by approximately fourteen countries, and several others are planning to reintroduce it as a main anti-vector tool; however, it strongly polarizes the opinion of scientists, who line up on the field as opponents, centrists or supporters, highlighting DDT health benefits or putative risks depending on their alignment (Bouwman et al., 2011). In this context, the present chapter will review the current knowledge on DDT use, and will suggest some possible future directions to be taken for malaria vector control.

The chapter will open on a short illustration of the *Plasmodium* life cycle, which occurs either in mosquito vector (sexual reproduction) or in human host (asexual replication). Since anti-vector control measures are directed to mosquito killing, *Plasmodium* sexual cycle will be prioritized. Therefore, the insecticides currently allowed for malaria vector control, including organochlorines (OCs), organophosphates (OPs), carbamates (Cs), and pyrethroids (PYs), will be briefly described. After such a brief introduction, a special attention will be paid to DDT. Formulation, cost-effectiveness, mechanisms of action, resistance and environmental issues will be discussed. The big debate among pro-DDT, DDT-centrist, or anti-DDT scientists will be examined. In this context, the state-of-the-art of knowledge on DDT toxicity will be analyzed, and few tips on possible alternatives to DDT will be given.

Taken altogether, these notions should help the reader to arise his own opinion on such a hot topic, in order to feed the ongoing debate. In areas endemic for malaria, is DDT dangerous as the bull in a China shop? Or perhaps is it worth using DDT, since its advantages related to malaria prevention are self-evident as the elephant in the room? Any answers aimed at finding the most practicable way to fight malaria through vector control are urgently required.

2. Materials and methods

All data were obtained from literature searches, by using the search engines Scopus and Pubmed. Because of the complexity of the subject, only the most relevant studies were selected, and reviews were prioritized. Old literature was accessed electronically, or hard copies were obtained from libraries. Information on human exposure and health effects was based on reviews published over the past ten years and supplemented with recent studies on exposure due to indoor spraying and treated bednets.

3. *Plasmodium* life cycle

Malaria parasites have evolved a complicated life cycle alternating between human and *Anopheles* mosquito hosts, as represented in Figure 1. Five *Plasmodium* strains (*P. falciparum*, *P. vivax*, *P. ovale*, *P. malariae*, and *P. knowlesi*) can affect humans in more than 90 countries, inhabited by 40% of the global population. In some of these areas, over 70% of residents are continuously infected by the most deadly form of the parasite, *P. falciparum*. Surviving children develop various levels of natural immunity; however, it does not protect them from repeated infections and illness throughout life.

3.1. *Plasmodium* life cycle in *Anopheles* mosquitoes

Plasmodium is transmitted to humans by female mosquitoes of *Anopheles* species. There are approximately 484 recognised species, and over 100 can transmit human malaria; however, only 30–40 commonly transmit *Plasmodium* parasites in endemic areas. *Anopheles gambiae* is one of the best known malaria vectors that lives in areas near human habitation (Rogier &

Hommel, 2011). The intensity of malaria parasite transmission varies geographically according to vector species of *Anopheles* mosquitoes. Risk is measured in terms of exposure to infective mosquitoes, with the heaviest annual transmission intensity ranging from 200 to >1000 infective bites per person. Interruption of transmission is technically difficult in many parts of the world because of limitations in approaches and tools for malaria control. In addition to ecological and behavioral parameters affecting vectorial capacity, *Anopheles* species also vary in their innate ability to support malaria parasite development. Environmental conditions such as temperature in mosquito microhabitats serve to regulate both the probability and timing of sporogonic development (Rogier & Hommel, 2011).

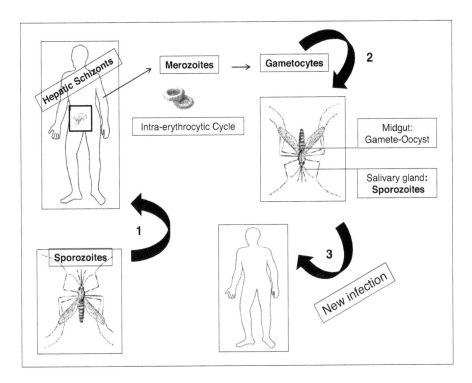

Figure 1. *Plasmodium* parasite life cycle.

In the mosquito, three phases of life of the parasite involve developmental transitions between gametocyte and ookinete stages, between ookinetes and mature oocysts, and between oocysts and sporozoites. When a female Anopheles sucks the blood of a malaria patient, the gametocytes also enter along with blood. They reach the stomach, and gamete formation takes place (Aly et al., 2009). Two types of gametes are formed: the microgametocytes (male) originate active microgametes, and the megagametocyte (female) undergoes some reorganization forming megagametes. Fertilization of the female gamete by the male gamete occurs rapidly after gametogenesis. The fertilization event produces a zygote that remains inactive for some time and then elongates into a worm-like ookinete. The ookinete is one of the most important stages of *Plasmodium* development in the mosquito. It is morphologically and biochemically distinct from the earlier sexual stages (gametocytes and zygote), and from the later stages (oocyst and sporozoites). Development to ookinete allows the parasite to escape from the tightly packed blood bolus, to cross the sturdy peritrophic matrix, to be protected from the digestive environment of the midgut lumen, and to invade the gut epithelium. The success of each of these activities may depend on the degree of the biochemical and physical barriers in the mosquito (such as density of blood bolus, thickness of peritrophic matrix, proteolytic activities in the gut lumen etc.) and the ability of the ookinete to overcome these barriers. Ookinete motility, resistance to the digestive enzymes, and recognition/invasion of the midgut epithelium may play crucial roles in the transformation to oocyst. At the end of the process oocysts produce sporozoites, which can navigate successfully to the salivary glands, where they will be ready for further infection of human beings, and continuation of their life cycle (Beier, 1998).

3.2. *Plasmodium* life cycle in humans

The transmission of the parasite to humans starts when the mosquito injects a small amount of saliva containing 5-200 sporozoites (resident in the salivary gland of the vector) into the skin of the human vector (Menard, 2005). Once in the bloodstream, sporozoites reach the liver and infect the hepatocytes (Trieu et al., 2006). In the liver district, sporozoites grow and change into a new structure of parasite called schizont, a large round cell. The schizont divides through an asexual reproduction (schizogony) resulting in the formation of a thousand small cells called merozoites. After a developmental period in liver, during which patients do not show any clinical symptoms of disease, merozoites are released from liver schizonts into the blood, entering host erythrocytes and starting the intraerythrocytic stage of parasite development (Banting et al., 1995).

This occurs inside a parasitophorous vacuole, the membrane of which separates the cytosol of the erythrocyte from the plasma membrane of the parasite. In the erythrocyte young 'ring' forms of the parasite grow to become trophozoites. Intraerythrocytic development is completed by the formation of new plasma membranes after multiple nuclear divisions (schizogony). Infectious merozoites are then released from the erythrocyte and a new cycle restart (Cowman & Crabb, 2006). One erythrocytic cycle is completed in 48 hours. The toxins are liberated into the blood along with merozoites. The toxins are then deposits in the liver, spleen and under the skin. The accumulated toxins cause malaria fever that lasts for 6 to 10 hours and then it comes again after every 48 hours with the liberation of new generated merozoites. During the erythrocytic stage, some merozoites increase in size to form two types of gametocytes, the macrogametocytes (female) and microgametocytes (male). This process is called gametocytogenesis. The specific causes underlying this sexual differentiation are largely unknown. These gametocytes take roughly 8–10 days to reach full maturity. The gametocytes develop only in the appropriate species of mosquito. If this does not happen, they degenerate and die (Rogier & Hommel, 2011).

4. Vector control as a key strategical approach for malaria eradication

The historical successful elimination of malaria in various parts of the world has been achieved mainly by vector control (Harrison, 1978). Since early nineteenth century (Breman, 2001), vector control has remained the most generally effective measure to prevent malaria transmission and therefore is one of the four basic technical elements of the Global Malaria Control Strategy. The principal objective of vector control is the reduction of malaria morbidity and mortality by reducing the levels of transmission. Vector control methods vary considerably in their applicability, cost and sustainability of their results.

4.1. Classification of insecticides used for vector control

The most prominent classes of insecticides act by poisoning the nervous system of insects, which is very similar to that of mammals. They are often subclassified by chemical type as organochlorines (OCs), organophosphates (OPs), carbamates (Cs) and pyrethroids (PYs) (Prato et al., 2012).

OCs belong to a larger class of compounds called chlorinated hydrocarbons, containing chlorine and including DDT. They have various chemical structures, and are cheap and effective against target species. OCs can alter and disrupt the movement of ions (calcium, chloride, sodium and potassium) into and out of nerve cells, but they may also affect the nervous system in other ways depending on their structure. OCs are very stable, slow to degrade in the environment and soluble in fats: unfortunately, due to persistence and fat solubility, OCs can bioaccumulate in the fat of large animals and humans by passing up the food chain.

OPs were developed in the 1940s as highly toxic biological warfare agents (nerve gases). On the other hand, Cs feature the carbamate ester functional group. OPs and Cs are very different at a chemical level; however, they have a similar mechanism of action. OPs and Cs block a specific enzyme, the acetylcholinesterase, which is able to remove an important neurotransmitter, the acetylcholine, from the area around the nerve cells stopping their communication. Hence, these insecticides are called acetylcholinesterase inhibitors. Structural differences between the various OPs and Cs affect the efficiency and degree of acetylcholinesterase blockage, highly efficient and permanent for nerve gases, temporary for commonly used pesticides. Many different OPs have been developed in order to replace DDT and find compounds that would be less toxic to mammals. Unfortunately, OP Parathion acute toxicity is greater than DDT, and this characteristic causes a significant number of human deaths.

Finally, synthetic PYs, developed in the 1980s, represent one of the newer classes of insecticides. Although their chemical structure is quite different from that of other insecticides, the target of action is also the nervous system. PYs affect the movement of sodium ions (Na^+) into and out of nerve cells that become hypersensitive to neurotransmitters.

4.2. Indoor Residual Spraying (IRS) and Insecticide-Treated Nets (ITNs)

Indoor residual spraying (IRS) with insecticides continues to be the mainstay for malaria control and represents the process of spraying stable formulations of insecticides on the inside walls of certain types of dwellings, those with walls made from porous materials such as mud or wood but not plaster as in city dwellings. Mosquitoes are killed or repelled by the spray, preventing the transmission of the disease. The main purpose of IRS is to reduce malaria transmission by reducing the survival of malaria vectors, life span of female mosquitoes, thereby reducing density of mosquitoes (WHO, 2006b). Several pesticides have historically been used for IRS, the first and most well-known being DDT.

Space spraying, or fogging, relies on the production of a large number of small insecticidal droplets, that resemble smoke or fog by rapidly heating the liquid chemical, intended to be distributed through a volume of air over a given period of time. When these droplets impact on a target insect, they deliver a lethal dose of insecticide. It is primarily reserved for application during emergency situations to rapidly reduce the population of flying insects in a specific area resulting in decrease of transmission (CDC, 2009). It is effective as a contact poison with no residual effect, thus it must be repeated at intervals of 5-7 days in order to be fully effective. The application must coincide with the peak activity of adult mosquitoes, because resting mosquitoes are often found in areas that are out of reach to the applied insecticides. The best moment to kill adult mosquitoes by fogging is at dusk, when they are most active in forming swarms. The most commonly used products are natural pyrethrum extract, synthetic PYs, and Malathion.

Mosquito nets treated with insecticides—known as insecticide treated nets (ITNs) or bednets—were developed in the 1980s for malaria prevention. Properly used, a mosquito net effectively offers protection against mosquitoes and other insects, and thus against the diseases they may carry. Two categories of ITNs are available: conventionally treated nets and long-lasting ITNs (LLINs). ITNs are estimated to be twice as effective as untreated nets, and offer greater than 70% protection compared with no net. These nets are impregnated with PYs, which will double the protection over a non-treated net by killing and repelling mosquitoes, and are proved to be a cost-effective prevention method against malaria (D'Alessandro et al., 1995). Washing and the associated regular retreatment of the nets determine a rapid loss of efficacy of ITNs, thus limiting the operational effectiveness of an ITN program (Lines, 1996).

Biological activity of LLINs, a relatively new technology, generally retains the efficacy for at least 3 years (WHO, 2005), and can reduce human–mosquito contact, which results in lower sporozoite and parasite rates. Different types of long-lasting insecticide impregnated materials are under field trials in different countries. Treatments of screens, curtains, canvas tents, plastic sheet, tarpaulin, etc., with insecticides may provide a cheap and practical solution for malaria vector control. Particularly, the residual insecticides in insecticide-treated wall lining (ITWL) are durable and can maintain control of insects significantly longer than IRS by providing an effective alternative or additional vector control tool (Munga et al., 2009).

5. Dichlorodiphenyltrichloroethane (DDT)

DDT is an OC insecticide; it is white, crystalline solid, tasteless, and almost odorless (PAN, 2012). It is a highly hydrophobic molecule, nearly insoluble in water but with good solubility in most organic solvents, such as fats and oils. DDT is not present naturally, but is produced by the reaction of chloral (CCl_3CHO) with chlorobenzene (C_6H_5Cl) in the presence of sulfuric acid, which acts as a catalyst. DDT was originally synthesised in 1874, but its action as an insecticide was not discovered until 1939. It was the first widely used synthetic pesticide, employed extensively by allied forces during the Second World War for the protection of military personnel from malaria and typhus, released commercially only in 1945. The Swiss chemist Paul Hermann Müller was awarded the Nobel Prize in Physiology or Medicine in 1948 for his discovery of the high efficiency of DDT as a contact poison against several arthropods.

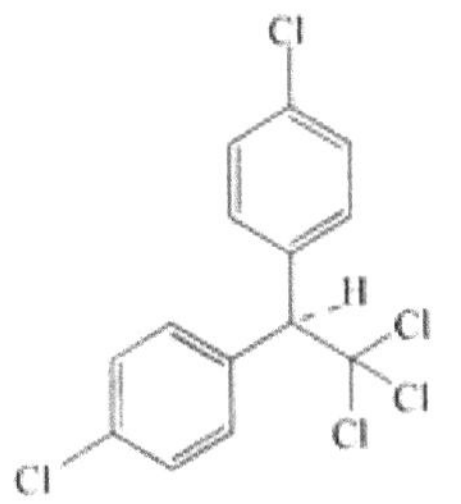

Figure 2. DDT

5.1. Production and use

While the post-war period also saw the introduction of most of the other major families of insecticides still in use today, DDT remained the most extensively used insecticide throughout the world until the mid 1960s. By this time, it had been credited with a number of significant public health successes, including the eradication of malaria from the United States and Europe (Attaran & Maharaj, 2000). DDT is currently being produced in three countries: India, China, and the Democratic People's Republic of Korea (North Korea). By far the largest amounts are produced in India for the purpose of disease vector control. In China, the average annual production during the period 2000–2004 was 4,500 metric tons of DDT, but 80–90% was used in the production of Dicofol, an acaricide, and around 4% was used as additive in antifouling paints. The remainder was meant for malaria control and was exported (PAN, 2012).

5.2. Cost-effectiveness

Both the effectiveness and costs of DDT are dependent on local settings and deserve careful consideration in relation to alternative products or methods (Walker, 2000). DDT has been known as the only insecticide that can be used as a single application in areas where the malaria transmission season is > 6 months. However, information is lacking on the potential variability in residual action of DDT (e.g., due to sprayable surface, climatic conditions, social factors). Direct costs of IRS are the procurement and transport of insecticide, training of staff, operations, awareness-raising of communities, safety measures, monitoring of efficacy and insecticide resistance, monitoring of adverse effects on health and the environment, and storage and disposal. Apart from the direct costs, it is essential that the unintended costs of DDT to human health and the environment are included in the cost assessment. In addition, contamination of food crops with DDT could negatively affect food export. A comprehensive cost assessment of DDT versus its alternatives should include the potential costs of atmospheric transport and chronic health effects.

5.3. Mechanism of action

The basic mechanism of action for most pesticides is an alteration in the transfer of a signal along a nerve fiber and across the synapse from one nerve to another or from a nerve to a muscle fiber. The transfer of a signal along a nerve occurs by changes in the electrical potential across the nerve cell membrane which is created by the movement of ions in and out of the cell. At the terminal end of a nerve, the signal is transferred across the synapse to the next nerve cell by the release of neurotransmitters. Different classes of pesticides inhibit this process in different ways, but the end result is an alteration in normal nerve signal propagation. OCs pesticides act primarily by altering the movement of ions across the nerve cell membranes, thus changing the ability of the nerve to fire.

The WHO has designated DDT as a Class II pesticide, based on its LD_{50} of 250 mg/kg (WHO, 1996). The mechanism by which DDT causes neurotoxicity is well studied. In insects DTT opens sodium ion channels in neurons, causing them to burn spontaneously. By causing repetitive firing of nerve cells, the cells eventually are unable to fire in response to a signal. DDT produces tremors and incoordination at low doses, convulsions at higher doses caused by the repetitive discharge (over-firing) of the nerves. Effects of chronic exposures to DDT are difficult to identify because they are general nervous systems alterations that can occur through many causes (apathy, headache, emotional lability, depression, confusion and irritability).

5.4. Resistance issues

As the number and size of programs that use DDT for indoor spraying increase, insecticide resistance is a matter of growing concern. Insects with certain mutations in their sodium channel gene are resistant to DDT and other similar insecticides. DDT resistance is also conferred by up-regulation of genes expressing cytochrome P450 in some insect species (Denholm et al., 2002).

Many insect species have developed resistance to DDT. The first cases of resistant flies were known to scientists as early as 1947, although this was not widely reported at the time (Metcalf, 1989). Since the introduction of DDT for mosquito control, DDT resistance at various levels has been reported from > 50 species of anopheline mosquitoes, including many vectors of malaria (Hemingway & Ranson, 2000). Unless due attention is paid to the role of insecticide resistance in the breakdown of the malaria eradication campaign of the 1960s, resistance may once again undermine malaria control.

In the past, the use of DDT in agriculture was considered a major cause of DDT resistance in malaria vectors, as many vectors breed in agricultural environments. By 1984 a world survey showed that 233 species, mostly insects, were resistant to DDT (Metcalf, 1989). Today, with cross resistance to several insecticides, it is difficult to obtain accurate figures on the situation regarding the number of pest species resistant to DDT. At present, DDT resistance is thought to be triggered further by the use of synthetic PYs (Diabate et al., 2002). This is due to a mechanism of cross-resistance between PYs and DDT, the so-called sodium channel mutation affecting neuronal signal transmission, which is governed by the *kdr* (knock-down resistance) gene (Martinez-Torres et al., 1998). The *kdr* gene is being reported from an increasing number of countries; thus, even in countries without a history of DDT use, resistance to DDT is emerging in populations of malaria vectors (WHO, 2006a). Contemporary data from sentinel sites in Africa indicate that the occurrence of resistance to DDT is widespread, especially in West and Central Africa. In Asia, the resistance to DDT is particularly widespread in India.

5.5. Environmental issues

Part of the success of DDT can be attributed to its persistence in the environment, thus reducing the need for frequent application. DDT is one of nine persistent organic pollutants (POPs) which bioaccumulate and are transported by air and water currents from warmer climates to temperate zones, where they have never been used. DDT has low to very low rates of metabolism and disposition, depending on ambient temperatures: the process of degradation is dramatically slowed down in cooler climates. It is degraded slowly into its main metabolic products, 1,1-dichloro-2,2-bis(p-chlorophenyl) ethylene (DDE) and dichloro-diphenyldichloroethane (DDD), which have similar physicochemical properties but differ in biological activity.

DDT is emitted through volatilization and runoff. It is more volatile in warmer than in colder parts of the world, which through long-range atmospheric transport results in a net deposition and thus gradual accumulation at high latitudes and altitudes (Harrad, 2001). Loss through runoff is low because DDT is extremely hydrophobic and has a strong affinity for organic matter in soils and aquatic sediment but is virtually insoluble in water. However, when applied to aquatic ecosystems, DTT is quickly absorbed by organisms and by soil or it evaporates, leaving little amount of DDT dissolved in the water itself (Agency for Toxic Substances and Disease Registry, 2002). Half-lives of DDT have been reported in the range of 3–7 months in tropical soils (Varca & Magallona 1994; Wandiga, 2001) and up to 15 years in temperate soils (Ritter et al., 1995). The half-life of each of its metabolic products is similar or longer.

The global risk of adverse effects to human health and the environment has led the international community to mandate the UN Environment Programme (UNEP) to convene an intergovernmental negotiating committee (INC) for a POPs Convention to phase out production and use (UNEP, 1997a; UNEP, 1997b). As a result of these environmental concerns, the use of DDT was increasingly restricted or banned in most developed countries after 1970.

DDT and its metabolic products present in the global environment have originated mostly from its previous large-scale use in agriculture and domestic hygiene. Because DDT is currently allowed only for indoor spraying for disease vector control, its use is much smaller than in the past. Nevertheless, DDT sprayed indoors may end up in the environment (e.g., when mud blocks of abandoned houses are dissolved in the rain). Even today, DDT remains so widespread in the environment that it is likely that exposure to it is unavoidable. While exposure in the industrialised world has fallen dramatically, exposure remains high in some developing countries where DDT continues to be used in vector control.

DDT is very fat-soluble and could therefore be found in fatty foods such as meat and diary products. Even in countries across North America and Northern Europe, where its use has been banned for over a decade DDT residues are still often found in food. This is because of environmental persistence, illegal use, or importation of contaminated food from regions where DDT is still used.

6. The big debate on DDT as anti-malaria tool

In 1955, the WHO commenced a program to eradicate malaria worldwide, relying largely on DDT. The program was initially very successful, eliminating the disease in Taiwan, much of the Caribbean, the Balkans, parts of northern Africa, the northern region of Australia, and a large swath of the South Pacific and dramatically reducing mortality in Sri Lanka and India (Harrison, 1978). However, widespread agricultural use led to resistant insect populations. In many areas, early victories partially or completely reversed, and in some cases rates of transmission even increased (Chapin & Wasserstrom,1981). The program was successful in eliminating malaria only in areas with "high socio-economic status, well-organized health-care systems, and relatively less intensive or seasonal malaria transmission" (Sadasivaiah et al., 2007). In tropical regions, DDT was less effective due to the continuous life cycle of mosquitoes and poor infrastructure. It was not applied at all in sub-Saharan Africa due to these perceived difficulties.

Moreover, the adverse health effects of DDT versus the health gains in terms of malaria prevention require more attention. For example, a gain in infant survival resulting from malaria control could be partly offset by an increase in preterm birth and decreased lactation, both of which are high risk factors for infant mortality in developing countries. The WHO is conducting a re-evaluation of health risks of DDT, but progress has been slow (PAN, 2012). Nevertheless, in 2006 it approved the use of DDT, particularly indoor residual spraying of walls, in areas endemic for malaria for health-related reasons (WHO, 2006a; WHO, 2006b), although it also carefully drew up major guidelines (WHO 2000). Currently, DDT represents one the main stays to achieve goals of Global Eradication Program launched in 2007 by the Bill and Melinda Gates Foundation, the World Health Organization (WHO) and the Roll Back Malaria association (Roberts & Enserink, 2007; Greenwood, 2008; Khadjavi et al., 2010; Prato et al., 2012). However, in the recent years the possible effects of DDT on human health have been a hot topic of discussion inside malaria research community, as certified by the large number of available publications and intense correspondence among scientists (e.g., Blair et al., 2009; Burton, 2009; van den Berg, 2009; Tren & Roberts, 2010; Bouwman et al., 2011; Tren & Roberts, 2011). The debate is heavily polarized, and three main viewpoints can be identified, as suggested by Bouwman et al. (Bouwman et al., 2011): anti-DDT, centrist-DDT, and pro-DDT.

6.1. Anti-DDT point of view

DDT opponents usually claim for DDT elimination because of environmental and health concerns. However, Tren & Roberts (Tren & Roberts, 2011) pointed that the "activist groups currently promote an anti-DDT agenda routinely hyping supposed human health and environmental harm from DDT and ignoring studies that find no association between DDT and such harm". As an example, Tren & Roberts mentioned the Biovision's "Stop DDT" project engaged to achieve a world-wide ban on DDT (Biovision, 2011), which apparently was connected to the Secretariat of the Stockholm Convention's promotion of an arbitrary deadline for cessation of DDT production by 2020 (United Nations Environment Programme, 2007). Another representative example of a recent anti-DDT action is given by a court case occurred in Uganda (Lewis, 2008): a petition filed in Kampala's High Court accused the Ugandan government of not following DDT spraying guidelines, whether those of the WHO or those of Uganda's National Environment Management Authority. In that case, it appears evident that the big matter was not DDT itself as a molecule, but its incorrect use. In this context, a major point questioned by anti-DDT scientists is that also IRS workers are highly exposed to DDT, since prescribed personal protection procedures and safe practices are not always followed, because of uncomfortable working conditions. Not wearing masks or gloves and frequent wiping of sweaty faces with the same cloth increases dermal and inhalation uptake leading to very high exposure (Bowman et al., 2011). Indeed, DDT serum levels in IRS workers in South Africa were high compared with the general population living in DDT-sprayed houses (Bouwman et al., 1991). On the other hand, Bimenya et al. (Bimeneya et al., 2010) did not found any DDT increase in serum of Ugandan DDT applicators over an entire spray season, stating that effective exposure reduction is possible when protective clothing is used and strict adherence to WHO guidelines (WHO, 2000) is observed. Nevertheless, the WHO's review of human health aspects of DDT use in IRS concluded that "for households where IRS is undertaken, there was a wide range of DDT and DDE serum levels between studies. Generally, these levels are below potential levels of concern for populations" (WHO, 2011b), and none of the thousands of studies conducted to find possible human health effects of

DDT satisfied even the most basic epidemiological criteria to prove a cause-and-effect relationship (Tren & Roberts, 2011).

6.2. Centrist-DDT point of view

According to Bouwman et al. (Bouwman et al., 2011), "the centrist-DDT point of view adopts an approach that pragmatically accepts the current need for DDT to combat malaria transmission using indoor residual spraying (IRS) but at the same time recognizes the risks inherent in using a toxic chemical in the immediate residential environment of millions of people". Thus, scientists sharing a centrist-DDT point of view such as Bouwman and colleagues suggest caution in using DDT because of insufficient investigation whether DDT is safe or not; however, they do recognize its undoubted benefits in areas endemic for malaria and its major role as a life-saving tool. In this context, DDT-centrists call for alternative chemicals, products, and strategies, eventually in order to terminate in the future any use of DDT in IRS for malaria control. As it will be discussed in paragraph 6, some vector control methods are already available as alternatives to DDT. Two of these, the use of alternative insecticides in IRS and the use of insecticide-treated bed nets (ITNs), are mainstreamed because of their proven impact on the malaria burden; other alternatives are receiving limited attention to date, but may play an important role in the future (van den Berg, 2009).

6.3. Pro-DDT point of view

DDT supporters consider DDT safe to use in IRS when applied correctly, and promote DDT to be used for IRS in malaria control where it is still effective. In their perspective, in a risk-benefit comparison, the eventual toxic effects of DDT would be far less than those caused by malaria (Africa Fighting Malaria, 2010; Roberts et al., 1997). Apparently, this is the point of view of WHO itself, since it approved in 2006 the use of DDT, particularly indoor residual spraying of walls, in areas endemic for malaria for health-related reasons (WHO, 2006a; WHO, 2006b), although it also carefully drew up major guidelines (WHO 2000). Moreover, several national malaria control programs and ministers of health repeatedly proclaimed the importance of DDT for disease control programs in countries with high incidence of malaria. These include Namibia and the Southern African Development Community (SADC), which recently reasserted that DDT is a major tool for malaria vector control and announced their intention to produce DDT locally (SADC, 2011). Similarly, the 35 heads of state of the countries members of the African Leaders Malaria Alliance (ALMA) recently endorsed use of DDT in indoor residual spraying (IRS) (ALMA 2010). As a matter of fact, as a consequence of the global eradication program recently launched by charity foundations, which invested relevant amounts of money in DDT-based vector control (Roberts & Enserink, 2007; Greenwood, 2008; Khadjavi et al., 2010; Prato et al., 2012), in 2010 World Health Organization (WHO) officially registered - for the first time in the last decade - a decline in estimated malaria cases and deaths, with 655.000 deaths counted among more than 200 million clinical cases worldwide (WHO 2011a).

7. Studies on DDT toxicity

Despite the concerns of DDT opponents (see par. 6.1), to date there is no consistent evidence that DDT or its metabolite DDE can be toxic for humans. Indeed, despite the large number of studies performed in this context, results are highly contradicting, probably due to different analytical conditions and approaches used by different researchers. On the other hand, DDT toxic effects on animals have been demonstrated quite convincingly. This should be taken in account in the context of general environmental issues (par. 5.5) which led to DDT ban in malaria-free countries. In the following sub-sections, current knowledge on DDT effects on animal and human health will be reviewed.

7.1. Animals

Due to its lipophilicity, DDT readily binds with fatty tissue in any living organism, and because of its chemical stability, bioconcentrates and biomagnifies with accumulation of DDT through the food chain, in particular in predatory animals at the top of the ecological pyramid (Jensen et al., 1969). By the mid 1950s, experimental studies on animals have demonstrated chronic effects on the nervous system, liver, kidneys, and immune systems in experimental animals attributable to DDT and DDE (Turusov et al., 2002), and it quickly became apparent that this could extend to the broader environment (Ramade, 1987). However, dose levels at which effects were observed are at very much higher levels than those which may be typically encountered in humans.

DDT is highly toxic to fish. The 96-hour LC50 (the concentration at which 50% of a test population die) ranges from 1.5 mg/litre for the largemouth bass to 56 mg/litre for guppy. Smaller fish are more susceptible than larger ones of the same species. An increase in temperature decreases the toxicity of DDT to fish (PAN, 2012).

DDT and its metabolites can lower the reproductive rate of birds by causing eggshell thinning which leads to egg breakage, causing embryo deaths. Sensitivity to DDT varies considerably according to species. Predatory birds and fish-eating birds at the top of the food chain are the most sensitive. The thickness of eggshells in peregrine falcons was found to have decreased dramatically following the pesticide's introduction (Ratcliffe, 1970), likely due to hormonal effects and changes in calcium metabolism (Peakall, 1969). Colonies of brown pelicans in southern California plummeted from 3000 breeding pairs in 1960 to only 300 pairs and 5 viable chicks in 1969. In the US, the bald eagle nearly became extinct because of environmental exposure to DDT. According to research by the World Wildlife Fund and the US EPA, birds in remote locations can be affected by DDT contamination. Albatross in the Midway islands of the mid-Pacific Ocean show classic signs of exposure to OCs chemicals, including deformed embryos, eggshell thinning and a 3% reduction in nest productivity. Researchers found levels of DDT in adults, chicks and eggs nearly as high as levels found in bald eagles from the North American Great Lakes (PAN, 1996).

7.1.1. Reproductive and teratogenic effects (birth defects)

DDT causes adverse reproductive and teratogenic effects in test animals. In one rat study, oral doses of 7.5 mg/kg/day for 36 weeks resulted in sterility. In rabbits, doses of 1 mg/kg/day administered on gestation days 4-7 resulted in decreased foetal weights. In mice, doses of 1.67 mg/kg/day resulted in decreased embryo implantation and irregularities in the oestrus cycle over 28 weeks (Agency for Toxic Substances and Disease Registry, 1994). Many of these observations may be the result of disruptions to the endocrine (hormonal) system.

In mice, maternal doses of 26 mg/kg/day DDT from gestation through to lactation resulted in impaired learning in maze tests.

7.1.2. Cancer

The evidence relating to DDT and carcinogenicity provides uncertain conclusions. It has increased tumour production, mainly in the liver and lungs, in test animals such as rats, mice and hamsters in some studies, but not in others. In rats, liver tumours were induced in three studies at doses of 12.5 mg/kg/day over periods of 78 weeks to life, and thyroid tumours were induced at doses of 85 mg/kg/day over 78 weeks. Tests have shown laboratory mice were more sensitive to DDT. Life time doses of 0.4 mg/kg/day resulted in lung tumours in the second generation and leukaemia in the third generation, and liver tumours were induced at oral doses of 0.26 mg/kg/day in two separate studies over several generations (PAN, 2012).

7.2. Humans

The US Department of Health and Human Services (DHHS) has determined that "DDT may reasonably be anticipated to be a human carcinogen". DHHS has not classified DDE and DDD, but the US Environmental Protection Agency (EPA) has stated that they are probable human carcinogens (PAN, 2012), suspecting DDT, DDD and DDE of being environmental endocrine disrupters (Colburn et al., 1996) which may affect human health. Based on the results of animal studies, DDT was suspected to cause cancer, diabetes, neurodevelopmental deficits, pregnancy and fertility loss (Beard, 2006). However, available epidemiological studies reject DDT contribution in the development of these diseases and results are still unclear (Beard, 2006).

7.2.1. Reproductive disorders

In vitro studies have shown DDT and its metabolites to have human estrogenic activity (Chen et al., 1997) and DDE to act as an androgen antagonist (Kelce et al., 1995). Some researchers have also hypothesized a trend for decreasing semen quality in the general human community following the introduction of DDT (Carlsen et al., 1992; Sharpe & Skakkebaek, 1993) suggesting that environmental exposure to OCs may be causing human endocrine disruption. However, the observed patterns may simply reflect geographic variations and lifestyle factors (Hauser et al., 2002).

Much of the epidemiologic research about the possible influence of pesticide exposure in general on pregnancy outcome suffers from significant methodological problems. The largest and most rigorous study of DDT and adverse reproductive outcomes was conducted in a US perinatal cohort of over 44,000 children born between 1959 and 1966 (Longnecker et al., 2001). DDE concentration was estimated in stored serum taken during pregnancy from mothers of 2380 children. Increasing concentrations of serum DDE were statistically and significantly related to preterm births, intra-uterine growth retardation (Siddiqui et al., 2003) and maternal diastolic blood pressure (Siddiqui et al., 2002). On the other hand, other studies have failed to find any relationship between maternal DDT exposure and birth weight (Gladen et al., 2003).

Both animal models and early human studies have suggested a link with exposure to the DDT and the most common adverse pregnancy outcome (spontaneous abortion) (Saxena et al., 1980). However, the results of recent research are inconsistent. One small case-control study nested in a longitudinal study of Chinese textile workers found significantly higher levels of DDE in women with spontaneous abortion than full term controls. (Korrick et al., 2001) On the other hand, other studies have been unable to find an association (Gerhard et al., 1998). Unclear findings have been identified about the impact of DDT on fertility (Cohn et al., 2003): the probability of daughters' pregnancy fell with increasing levels of DDT in maternal serum, but it increased with increasing levels of DDE. Finally, OCs appear to transfer freely across the placenta from mother to foetus and could be also excreted in human milk (PAN, 2012).

In the late 1960s, concentrations of DDE in animals and first-trimester human fetal tissues correlated with reproductive abnormalities in male offspring such as hypospadias and undescended testes (Gray et al., 2001). A case-control study nested in a US birth cohort (1959–1966) (Longnecker et al., 2002) showed small increases in crypt-orchidism, hypospadias, and polythelia among boys with the highest DDE maternal serum levels when compared with those with the lowest maternal levels, although none of these were statistically significant. On the other hand, other studies failed to find a significant association between influence of DDT exposure on hormone levels in adult men, or DDT levels and sperm concentration/mobility in male partners of sub-fertile couples (Hauser et al., 2003).

7.2.2. Other endocrine conditions

Bone mineral density, which is regulated by the antagonistic effect of androgens and oestrogens, may be another possible target of endocrine disruption. DDT has been shown to modulate trophoblast calcium handling functions *in vitro* (Derfoul et al., 2003) and two small cross-sectional studies have suggested there may be a weak association between serum DDE levels and reduced bone mineral density (Beard et al., 2000; Glynn et al., 2000). However, a third study failed to demonstrate any correlation (Bohannon et al., 2000).

In vitro studies suggest that DDT and its metabolites do not influence thyroid metabolism (Langer et al., 2003; Rathore et al., 2002). Other research has failed to find a significant association with endometriosis, a hormone dependant pelvic inflammatory disease (Lebel et al., 1998).

7.2.3. Cancer

Breast cancer has been studied most rigorously; even though the majority of results showed no causative association with DDT exposure (Beard et al., 2006), the latest evidence indicates an increased risk in women who were exposed at a young age. It was hypothesised that DDT co-genres and metabolites might act as tumour promoters in hormonally sensitive cancers due to their oestrogenic and anti-androgenic properties (Iscan et al., 2002). More recently, larger and better designed studies have generally not supported this hypothesis (Calle et al., 2002; Snedeker, 2001). Other hormonally sensitive cancers include cancer of the endometrium and prostate. Two case-control studies have explored the possibility that DDT may be related to endometrial cancer with neither finding a significant association (Sturgeon et al., 1998; Weiderpass et al., 2000). On the other hand, an Italian hospital-based multisite case-control study of prostate cancer found an increased risk among farmers exposed to DDT (Settimi et al., 2003), although exposure assessment in this study relied on self-report, leaving these findings susceptible to recall bias. Rates of prostate cancer were also found to be increased among male applicators using chlorinated pesticides in the Agricultural Health Study cohort (Alavanja et al., 2003) and in a Swedish cohort of pesticide applicators (Dich & Wiklund, 1998).

Pesticides have been associated with pancreatic cancer (Beard, 2006). A large Norwegian prospective study of lifestyle factors and pancreatic cancer identified a higher risk among men occupied in farming, agriculture or forestry (Nilsen & Vatten, 2000). Recent research lends a physiological plausibility to a possible association between DDT and pancreatic cancer by suggesting that DDT may modulate oncogene expression or provide a growth advantage to mutated cells, for example, through its actions as an endocrine disrupter (Porta et al., 1999).

Case control studies using self-reported exposure have found significant associations between DDT exposure and lung cancer, leukaemia and non-Hodgkins lymphoma (NHL) (Beard, 2006). However a nested case-control study using stored serum identified a dose response relationship for NHL with PCB exposure but not DDT. A small case-control study using serum levels drawn at diagnosis has suggested an association between DDT exposure and colorectal cancer.

7.2.4. Nervous system

Animal studies have suggested DDT may cause central nervous system (CNS) toxicity (Eriksson & Talts, 2000). Exposure to DDT may be associated with a permanent decline in neurobehavioral functioning and an increase in psychiatric symptoms, but the few studies and limited exposure information made it impossible to be confidant about this potential relationship (Colosio et al., 2003). These findings are also complicated by potential confounding from exposure to other pesticides, such as organophosphates, that are known to have neurological effects. One recent case study suggested that DDT may be related to neurological impairment (Hardell et al., 2002). Another recent study of retired malaria-control workers found various neurobehavioral functions and performance deteriorated significantly with increasing years of DDT application (van Wendel de Joode et al., 2001). Subjects exposed to pesticides including DDT also scored worse than non-exposed subjects on a self-reported neuropsychological questionnaire of surviving members of a historical cohort of pesticide applicators (Beard, 2006).

7.2.5. Immune system

At least one cross-sectional study has associated DDT and other pesticide exposures with suppression or induction of several immune parameters (Daniel et al., 2002).

7.2.6. Diabetes

Diabetes has been associated with OC exposure in at least one study. An Australian cohort study of mortality in staff working as part of an insecticide application program also found increased mortality from pancreatic cancer in DDT-exposed subjects and from diabetes in subjects working with any pesticide (Beard, 2006).

7.3. Epidemiological studies

It is only in the last 25 years that more rigorous epidemiological research has focused on the possible adverse effects of exposure to DDT in humans. Unfortunately, they are not easily answered since epidemiologic research in this field is plagued by methodological challenges (Blondell, 1990). Fewer early human studies have been undertaken specifically on DDT, moreover they were small and limited in scope. A major methodological challenge is the difficulty in getting accurate information on subject exposure since many of the possible adverse effects of DDT (for example, cancer) may not become evident until many years after a causative exposure. Moreover, since it is rare for past exposure to have been accurately recorded at the time, exposure estimation has often been based on the response by subjects to questioning. However, subjects may have been unaware of significant past exposures to DDT through the food chain and even occupationally exposed subjects are unlikely to accurately remember and quantify exposures faced 20–30 years in the past. In the absence of a recorded exposure history, biological sampling of subjects may give some measure of their past exposure. Unlike other pesticides, DDT and DDE are only very slowly eliminated, making biological monitoring a relatively accurate, easy and cheap means of assessing past exposure. Serum levels of DDT and DDE are closely correlated with levels in adipose tissue and thus provide a relatively non-invasive measure (Mussalo-Rauhamaa, 1991). Unfortunately, biological monitoring of DDT presents its own potential for epidemiological bias since levels can also be influenced by factors that relate directly to the outcome of interest, in particular weight change.

Since DDT and its metabolites are so persistent in the environment and human tissues, humans are not excluded from this ecological trends raising questions about the possible impact of widespread pesticide exposure on human communities. Biological sampling near the time of peak use during the 1960s showed increasing DDT levels in most human communities, mainly due to exposure to residues in food. High levels of human exposure to DDT among those living in sprayed houses, most of whom are living under conditions of poverty and often with high levels of immune impairment, have been found in studies in South Africa and Mexico (Aneck-Hahn et al., 2007; Bouwman et al., 1991; De Jager et al., 2006; Yanez et al., 2002), but contemporary peer-reviewed data from India, the largest consumer of DDT, are lacking. The simultaneous presence of, and possible interaction between, DDT, DDE and PYs in human tissue is another area of concern (Bouwman et al., 2006; Longnecker, 2005). In North America, rather high levels of exposure have been recorded in biological samples collected in the 1960s (Eskenazi et al., 2009). DDT accumulates in fatty tissue and is slowly released. The half-life of DDT in humans is > 4 years; the half-life for DDE is probably longer (Longnecker, 2005).

8. Possible alternatives to DDT

Several vector control methods are currently available as alternatives to DDT, while others are under development. As previously stated, the use of alternative insecticides in IRS and the use of insecticide-treated bed nets (ITNs), are mainstreamed because of their proven impact on the malaria burden. Moreover, several non-chemical approaches could play a pivotal role in the future. Table 1 summarizes some possible alternative methods to DDT.

Alternatives to DDT	Chemical (yes/no)	Vector stage	Availability	Delivery/Resources	Risk
attractants	yes	adult	under development	local, private sector	resistance, toxicity
botanicals	no	larva adult	available	local	toxicity
chemical larviciding	yes	larva	available	spray teams	resistance, effect on ecosystems
design of irrigation structures	no	larva	available	irrigation sector	negligible
elimination of breeding sites	no	larva	available	local	negligible
fungi	no	adult	under development	not applicable	negligible
genetic methods	no	adult	under development	not applicable	to be studied
habitat manipulation	no	larva	available	local, agricolture sector	negligible
house improvement	no	adult	available	local, development programs	resistance

Alternatives to DDT	Chemical (yes/no)	Vector stage	Availability	Delivery/Resources	Risk
indoor residual spraying	yes	adult	available	spray teams	resistance, toxicity
insecticide-treated bednets	yes	adult	available	free distribution, social marketing, private sector	resistance, toxicity
irrigation management	no	larva	available	local, irrigation sector	negligible
microbial larvicides	no	larva	available	programs, private sectors	resistance
polystyrene beads	no	larva	available	local	negligible
predation	no	larva	available	local, programs, agriculture sector	negligible
repellents	yes	adult	under development	local, private sector	resistance, toxicity

Table 1. Alternative methods for malaria vector control. Adapted from (van den Berg, 2009)

8.1. Chemical methods

The strength of IRS with insecticides lies in its effect on shortening the life span of adult mosquitoes near their human targets (MacDonald, 1957). Two new approaches are currently being developed with regard to IRS, including some existing insecticides not currently available for public health (chlorfenapyr and indoxacarb), potentially effective in areas with pyrethroid resistance (N'Guessan et al., 2007a; N'Guessan et al., 2007b), and new formulations of existing insecticides with prolonged residual activity (Hemingway et al., 2006).

The main alternative to IRS are ITNs, which have been shown convincingly to substantially reduce all-cause child mortality, under both experimental (Lengeler, 2004) and operational conditions (Schellenberg et al., 2001; Fegan et al., 2007). Various new developments in ITN technology have spread recently. At least one nonpyrethroid insecticide with novel chemistry has been developed for ITNs (Hemingway et al., 2006) to cope with the problem of resistance; however, safety issues are still a concern. Other new ITN products are not expected to come to market in the short term.

Chemical insecticides as larvicides can play an important role to control mosquito breeding in urban settings, but they are a concern to the integrity of aquatic ecosystems.

Moreover, in order to push away mosquitoes, which usually are attracted by the moisture, warmth, carbon dioxide or estrogens from human skin, a large spectrum of repellents have been developed and are currently used; these substances, manufactured in several forms, including aerosols, creams, lotions, suntan oils, grease sticks and cloth-impregnating laundry emulsions, are usually applied on the skin or clothes, and produce a vapor layer characterized by bad smell or taste to insects (Brown & Hebert, 1997). The ideal repellent should satisfy several criteria: a) have long-lasting effectiveness; b) do not irritate human skin; c) have a bad odor only to mosquitoes but not to people; d) have no effects on clothes; e) be inert to plastics commonly used, such as glasses or bracelets; f)be chemically stable; and g) be economical (Brown & Hebert, 1997). The list of main insect repellents, some of which are also used as insecticides, includes N,N-diethyl-3-methylbenzamide (DEET), permethrin, picaridin, indalone, and botanicals (Prato et al., 2012). Additionally, innovative work is in progress on the attractiveness of human odors to malaria vectors, with potential applications as mosquito attractants and repellents for use in trapping and personal protection (Zwiebel & Takken, 2004).

8.2. Nonchemical methods

The development of non-chemical strategies alternative to insecticides and repellents is already available or currently on study. Before the advent of synthetic insecticides, vector control depended primarily on environmental management, and a meta-analysis of data mostly from that period indicated that it substantially reduced malaria risk (Keiser et al., 2005).

Elimination of vector-breeding habitats and managements of water bodies plays a key role in vector suppression, (Walker & Lynch, 2007). In irrigated agriculture, vector breeding can be controlled, through land leveling and intermittent irrigation (Keiser et al., 2002).

The role of aquatic predators as control agents of malaria vectors is potentially enhanced through conservation or through the introduction of agents from outside. Larvivorous fish have frequently been reared and released for controlling vector breeding in small water tanks and wells, but successes have generally been limited to more or less permanent water bodies (Walker & Lynch, 2007).

Microbial larvicides such as *Bacillus thuringiensis israelensis* and *Bacillus sphaericus* produce mosquito-specific toxins associated with a low risk of resistance development (Lacey, 2007). Recent field trials and pilot projects have shown good potential of both bacteria to manage mosquito breeding and to reduce biting rates in certain settings (Fillinger et al., 2008).

Also, insect pathogenic fungi have shown promising results for controlling adult *Anopheles* mosquitoes when sprayed on indoor surfaces and have potential to substantially reduce malaria transmission (Scholte et al., 2005).

Novel methods under development are genetically engineered mosquitoes and the sterile insect technique (Catteruccia, 2007). Genetic control appears a promising tool, comprising all methods by which a mechanism for pest or vector control is introduced into a wild population through mating. These include the sterile insect release method or the sterile insect technique (SIT), through which males are sterilized by irradiation or other means and released to mate with wild females, leading them to lay sterile eggs. Additionally, the introduction of genetic factors into wild populations aimed to make pests harmless to humans might be relevant (Pates & Curtis, 2005).

Finally novel approaches against vector borne diseases include transgenesis and paratransgenesis to reduce vector competence (Coutinho-Abreu et al., 2010). For vector transgenesis, the goal is to transform vectors with a gene (or genes) whose protein(s) impair pathogen development. Several mosquito species vectors of different parasites and viruses have been transformed. Some of the transformed mosquitoes were shown capable of blocking pathogen development via tissue-specific expression of molecules impairing the pathogen attachment to the midgut (Ito et al., 2002), or activating some biochemical pathways detrimental to pathogen survival (Franz et al., 2006). Paratransgenesis aims to reduce vector competence by genetically manipulating symbionts. Transformed symbionts are spread maternally or via coprophagy across an insect population (Durvasula et al., 1997).

Unfortunately, although these approaches are potentially promising, they remain a complex approach with a limited use (Coutinho-Abreu et al., 2010). Also, data on the cost-effectiveness of nonchemical methods are scarce. In a retrospective analysis of data from Zambia, environmental management was as cost-effective as ITNs (Utzinger et al., 2001). Moreover, environmental management can benefit from local resources, reducing the need for external funds.

9. Conclusion

To date, DDT represents a major tool for vector control in areas endemic for malaria, and in 2010 it was the main stay contributing to reduce malaria burden. Despite the big ongoing debate whether improve or ban its use, no convincing evidence on long-term toxic effects of DDT on humans is currently available. In the future, further constructive research aimed at ascertaining DDT effects on human health will be certainly welcome; also, the concurrent use of safe DDT alternatives (as long as they are effective as DDT, of course), should not be neglected. Nevertheless, DDT benefits appear self-evident up to now, thereby justifying its current use as an effective anti-malaria tool.

Author details

Mauro Prato[1,2], Manuela Polimeni[1] and Giuliana Giribaldi[1]

1 Dipartimento di Genetica, Biologia e Biochimica, Facolta' di Medicina e Chirurgia, Universita' di Torino, Italy

2 Dipartimento di Neuroscienze, Facolta' di Medicina e Chirurgia, Universita' di Torino, Italy

References

[1] Africa Fighting Malaria (2010). Africa Fighting Malaria. Indoor Residual Spraying and DDT. Available from http://www.fightingmalaria.org/pdfs/Africa%20Fighting%20Malaria%20IRS%20DDT%20issues.pdf

[2] Agency for Toxic Substances and Disease Registry. (1994). *Toxicology Profile for 4,4'-DDT, 4,4'-DDE, 4,4'DDD*, Public Health Service, Atlanta, GA.

[3] Agency for Toxic Substances and Disease Registry. (September 2002). *Toxicological Profile: for DDT, DDE, and DDE*, U.S. DEPARTMENT OF HEALTH AND HUMAN SERVICES Public Health Service, Retrieved from http://www.atsdr.cdc.gov/toxprofiles/tp35.pdf

[4] Alavanja, M.C., Samanic, C., Dosemici, M., Lubin, J., Tarone, R., Lynch, C., Knott, C., Thomas, K., Hoppin, J.A., Barker, J., Coble, J., Sandler, D.P. & Blair, A. (2003). Use of agricultural pesticides and prostate cancer risk in the Agricultural Health Study cohort. *The American Journal of Epidemiology*, Vol.157, No.9, pp. 800–814, ISSN 1476-6256

[5] ALMA (African Leaders Malaria Alliance) (2010). Report to ALMA Heads of State and Government. ALMA.

[6] Aly, A.S., Vaughan, A.M. & Kappe, S.H. (2009). Malaria parasite development in the mosquito and infection of the mammalian host. *Annual Reviews of Microbiology*, Vol. 63, No.10, pp. 195-221, ISSN 0066-4227

[7] Aneck-Hahn, N.H., Schulenburg, G.W., Bornman, M.S., Farias, P. & De Jager, C. (2007). Impaired semen quality associated with environmental DDT exposure in young men living in a malaria area in the Limpopo Province, South Africa. *Journal of Andrology*, Vol.28, pp. 423–434, ISSN 1939-4640

[8] Arbuckle, T.E. & Sever, L.E. (1998). Pesticide exposures and fetal death: a review of the epidemiologic literature. *Critical Reviews in Toxicology*, Vol.28, pp. 229–270, ISSN 1547-6898

[9] Attaran, A. & Maharaj, R. (2000). Doctoring malaria, badly: the global campaign to ban DDT. *British Medical Journal*, Vol.321, pp. 1403–1405, ISSN 0007-1447

[10] Banting, G., Benting, J. & Lingelbach, K. (1995). A minimalist view of the secretory pathway in *Plasmodium falciparum*. *Trends in Cell Biology*, Vol.5, pp 340-343, ISSN 0962-8924

[11] Beard, J. (2006). DDT and human health. Science of the Total Environment, Vol.355, pp. 78– 89, ISSN

[12] Beard, J., Marshall, S., Jong, K., Newton, R., Triplett-McBride, T., Humphries, B. & Bronks, R. (2000). 1,1,1-Trichloro-2,2-bis (p-chlorophenylethane (DDT) and reduced bone mineral density. *Archives of Environmental Health*, Vol.55, pp. 177–180, ISSN 0003-9896

[13] Beier, J.C. (1998). Malaria parasite development in mosquitoes. *Annual Review of Entomology*, Vol.43, pp. 519-543, ISSN 0066-4170

[14] Bimenya, G.S., Harabulema, M., Okot, J.P., Francis, O., Lugemwa, M. & Okwi, A.L. (2010). Plasma levels of DDT/DDE and liver function in malaria control personnel 6

months after indoor residual spraying with DDT in Uganda, 2008. *South African Medical Journal*, Vol.100, No.2, pp.118–121, ISSN 0256-9574

[15] Biovision (2011). Projects: International. Stop DDT: Promoting Effective and Environmentally Sound Alternatives to DDT, Available from http://www.biovision.ch/fileadmin/pdf/e/projects/2011/International/BV_Factsheet_Stop_DDT_2011_E.pdf

[16] Blair, A., Saracci, R., Vineis, P., Cocco, P., Forastiere, F., Grandjean, P., Kogevinas, M., Kriebel, D., McMichael, A., Pearce, N., Porta, M., Samet, J., Sandler, D.P., Costantini A.S. & Vainio, H. (2009). Epidemiology, public health, and the rhetoric of false positives. *Environmental Health Perspectives*, Vol.117, No.12, pp. 1809–1813, ISSN 0091-6765

[17] Blondell, J. (1990). Problems encountered in the design of epidemiologic studies of cancer in pesticide users. *La Medicina del Lavoro*, Vol.81, pp. 524–529, ISSN 0025-7818

[18] Bohannon, A.D., Cooper, G.S., Wolff, M.S. & Meier, D.E. (2000). Exposure to 1,1-dichloro-2,2-bis(p-chlorophenyl)ethylene (DDT) in relation to bone mineral density and rate of bone loss in menopausal women. *Archives of Environmental Health*, Vol.55, pp. 386–391, ISSN 0003-9896

[19] Bouwman, H., Cooppan, R.M., Botha, M.J. & Becker P.J. (1991). Serum levels of DDT and liver function of malaria control personnel. *South African Medical Journal*, Vol.79, No.6, pp.362–329, ISSN 0256-9574

[20] Bouwman, H., van den Berg, H. & Kylin, H. (2011). DDT and Malaria Prevention: Addressing the Paradox. *Environmental Health Perspectives*, Vol.119, No.6, pp. 744-747, ISSN 0091-6765

[21] Bouwman, H., Cooppan, R.M., Becker, P.J. & Ngxongo, S. (1991). Malaria control and levels of DDT in serum of two populations in Kwazulu. *Journal of Toxicology and Environmental Health*, Vol.33, pp. 141–155, ISSN 1087-2620

[22] Bouwman, H., Sereda, B. & Meinhardt, H.M. (2006). Simultaneous presence of DDT and pyrethroid residues in human breast milk from a malaria endemic area in South Africa. *Environmental Pollution*, Vol.144, pp. 902–917, ISSN 0269-7491

[23] Breman, J.G. (2001). The ears of the hippopotamus: manifestations, determinants and estimates of the malaria burden. *American Journal of Tropical Medicine and Hygiene*, Vol.64, No.1-2Suppl, pp. 1–11, ISSN 0002-9637

[24] Brown, M. & Herbert, A.A. (1997). Insect repellents: An overview. *Journal of the American Academy of Dermatology*, Vol.36, No.2Pt1, pp. 243-249, ISSN 1097-6787

[25] Burton, A. (2009). Toward DDT-free malaria control. *Environmental Health Perspectives*, Vol.117, No.8, pp. A344, ISSN 0091-6765

[26] Calle, E.E., Frumkin, H., Henley, S.J., Savitz, D.A. & Thun, M.J. (2002). Organochlorines and breast cancer risk. *CA a cancer journal for clinicians*, Vol.52, pp. 301–309, ISSN 1542-4863

[27] Carlsen, E., Giwercman, A., Keiding, N. & Skakkebaek, N. (1992). Evidence for decreasing quality of semen during past 50 years. *British Medical Journal*, Vol.305, pp. 609–612, ISSN 0007-1447

[28] Catteruccia, F. (2007). Malaria vector control in the third millennium: progress and perspectives of molecular approaches. *Pest Management Science*, Vol.63, pp. 634–640, ISSN 1526-4998

[29] CDC. (2009). Malaria - Vector Control. In: Center for disease control and prevention, 21 April 2009, Available from http://www.cdc.gov/malaria/ control_prevention/ vector_control.htm.

[30] Chapin, G. & Wasserstrom, R. (1981). Agricultural production and malaria resurgence in Central America and India. *Nature*, Vol.293, No.5829, pp. 181-185, ISSN 0028-0836

[31] Chen, C., Hurd, C., Vorojeikina, D., Arnold, S. & Notides, A. (1997). Transcriptional activation of the human estrogen receptor by DDT isomers and metabolites in yeast and MCF-7 cells. *Biochemical Pharmacology*, Vol.53, pp. 1161–1172, ISSN 0006-2952

[32] Cohn, B.A., Cirillo, P.M., Wolff, M.S., Schwingl, P.J., Cohen, R.D., Sholtz, R.I., Ferrara, A., Christianson, R.E., van den Berg, B.J. & Siiteri, P.K. (2003). DDT and DDE exposure in mothers and time to pregnancy in daughters. *The Lancet*, Vol.361, pp. 2205–2206, ISSN 1474-547X [erratum appears in Lancet 2003, Vol.362, No.9394, pp. 1504]

[33] Colburn, T., Dumanoski, D. & Meyers, J.P. (1996). Our Stolen Future, Penguin Books, New York, US.

[34] Colosio, C., Tiramani, M. & Maroni, M. (2003). Neurobehavioral effects of pesticides: state of the art. *Neurotoxicology*, Vol.24, pp. 577–591, ISSN 0161-813X

[35] Coutinho-Abreu, I.V., Zhu, K.Y. & Ramalho-Ortigao, M. (2010). Transgenesis and paratransgenesis to control insect-borne diseases: Current status and future challenges. *Parasitology International*, Vol.59, No1, pp. 1–8, ISSN 1873-0329

[36] Cowman, A.F. & Crabb, B.S. (2006). Invasion of Red Blood Cells by Malaria Parasites. Cell, Vol.124, No.4, pp. 755–766, ISSN 0092-8674

[37] D'Alessandro, U., Olaleye, B.O., McGuire, W., Thomson, M.C., Langerock, P., Bennett, S. & Greenwood, B.M. (1995). A comparison of the efficacy of insecticide-treated and untreated bed nets in preventing malaria in Gambian children. *Transactions of the Royal Society of Tropical Medicine and Hygiene*, Vol.89, No.6, pp. 596–598, ISSN 1878-3503

[38] Daniel, V., Huber, W., Bauer, K., Suesal, C., Conradt, C. & Opelz, G. (2002). Associations of dichlorodiphenyltrichloroethane (DDT) 4.4 and dichlorodiphenyldichloroethylene (DDE) 4.4 blood levels with plasma IL-4. *Archives of Environmental Health*, Vol.57, pp. 541–547, ISSN 0003-9896

[39] De Jager, C., Farias, P., Barraza-Villarreal, A., Avila, M.H., Ayotte, P., Dewailly, E., Dombrowski, C., Rousseau, F., Sanchez, V.D. & Bailey, J.L. (2006). Reduced seminal parameters associated with environmental DDT exposure and p,p'-DDE concentrations in men in Chiapas, Mexico: a cross-sectional study. *Journal of Andrology*, Vol.27, pp. 16–27, ISSN 1939-4640

[40] Denholm, I., Devine, G.J. & Williamson, M.S. (2002). Evolutionary genetics. Insecticide resistance on the move. *Science*, Vol.297, No.5590, pp. 2222–2223, ISSN 1095-9203

[41] Derfoul, A., Lin, F.J., Awumey, E.M., Kolodzeski, T., Hall, D.J. & Tuan, R.S. (2003). Estrogenic endocrine disruptive components interfere with calcium handling and differentiation of human trophoblast cells. *Journal of Cellular Biochemistry*, Vol.89, pp. 755–770, ISSN 1097-4644

[42] Diabate, A., Baldet, T., Chandre, F., Akoobeto, M., Guiguemde, T.R., Darriet, F., Brengues, C., Guillet, P., Hemingway, J., Small, G.J. & Hougard, J.M. (2002). The role of agricultural use of insecticides in resistance to pyrethroids in Anopheles gambiae s.l. in Burkina Faso. *American Journal of Tropical Medicine and Hygiene*, Vol.67, No.6, pp. 617–622, ISSN 0002-9637

[43] Dich, J. & Wiklund, K. (1998). Prostate cancer in pesticide applicators in Swedish agriculture. Prostate, Vol.34, pp. 100–112, ISSN

[44] Durvasula, R.V., Gumbs, A., Panackal, A., Kruglov, O., Aksoy, S., Merrifield, R.B., Richards, F.F. & Beard, C.B. (1997). Prevention of insect-borne disease: an approach using transgenic symbiotic bacteria. *Proceedings of the National Academy of Sciences of the United States of America*, Vol.94, No.7, pp. 3274–3278, ISSN 1091-6490

[45] Eriksson, P. & Talts, U. (2000). Neonatal exposure to neurotoxic pesticides increases adult susceptibility: a review of current findings. *Neurotoxicology*, Vol.21, pp. 37–47, ISSN 0161-813X

[46] Eskenazi, B., Chevrier, J., Rosas, L.G., Anderson, H.A., Bornman, M.S., Bouwman, H., Chen, A., Cohn, B.A., de Jager, C., Henshel, D.S., Leipzig, F., Leipzig, J.S., Lorenz, E.C., Snedeker, S.M. & Stapleton, D. (2009). The Pine River statement: human health consequences of DDT use. *Environmental Health Perspectives*, Vol.117, pp. 1359–1367, ISSN 0091-6765

[47] Fegan, G.W., Noor, A.M., Akhwale, W.S., Cousens, S. & Snow, R.W. (2007). Effect of expanded insecticide-treated bednet coverage on child survival in rural Kenya: a longitudinal study. *The Lancet*, Vol.370, pp. 1035–1039, ISSN 0140-6736

[48] Fillinger, U., Kannady, K., William, G., Vanek, M.J., Dongus, S., Nyika, D., Geissbühler, Y., Chaki, P.P., Govella, N.J., Mathenge, E.M., Singer, B.H., Mshinda, H., Lindsay, S.W., Tanner, M., Mtasiwa, D., de Castro, M.C. & Killeen, G.F. (2008). A tool box for operational mosquito larval control: preliminary results and early lessons from the Urban Malaria Control Programme in Dar es Salaam, Tanzania. *Malaria Journal*, Vol. 7, pp. 20, ISSN 1475-2875

[49] Franz, A.W., Sanchez-Vargas, I., Adelman, Z.N., Blair, C.D., Beaty, B.J., James, A.A. & Olson, K.E. (2006). Engineering RNA interference-based resistance to dengue virus type 2 in genetically modified Aedes aegypti. *Proceedings of the National Academy of Sciences of the United States of America*, Vol.103, No11, pp. 4198–203, ISSN 1091-6490

[50] Gerhard, I., Daniel, V., Link, S., Monga, B. & Runnebaum, B. (1998). Chlorinated hydrocarbons in women with repeated miscarriages. *Environmental Health Perspectives*, Vol.106, pp. 675–681, ISSN 0091-6765

[51] Gladen, B.C., Shkiryak-Nyzhnyk, Z.A., Chyslovska, N., Zadorozhnaja, T.D. & Little, R.E. (2003). Persistent organochlorine compounds and birth weight. *Annals of Epidemiology*, Vol.13, pp. 151–157, ISSN 1047-2797

[52] Glynn, A.W., Michaelsson, K., Lind, P.M., Wolk, A., Aune, M., Atuma, S., Darnerud, P.O. & Mallmin, H. (2000). Organochlorines and bone mineral density in Swedish men from the general population. *Osteoporosis International*, Vol.11, pp. 1036–1042, ISSN 1433-2965

[53] Gray, L.E., Ostby, J., Furr, J., Wolf, C.J., Lambright, C., Parks, L., Veeramachaneni, D.N., Wilson, V., Price, M., Hotchkiss, A., Orlando, E. & Guillette, L. (2001). Effects of environmental antiandrogens on reproductive development in experimental animals. *Human Reproduction Update*, Vol.7, pp. 248–264, ISSN 1460-2369

[54] Greenwood, BM. (2008). Control to elimination: implications for malaria research. *Trends in Parasitology*, Vol.24, No.10, (October 2008), pp. 449-454, ISSN 1471-5007

[55] Hardell, L., Lindstrom, G. & Van Bavel, B. (2002). Is DDT exposure during fetal period and breast-feeding associated with neurological impairment? *Environmental Research*, Vol.88, pp. 141–144, ISSN 0013-9351

[56] Harrad, S. (2001). Persistent Organic Pollutants: Environmental Behaviour and Pathways of Human Exposure. Dordrecht, the Netherlands:Kluwer Academic Publishers

[57] Harrison, G. (1978). Mosquitoes, Malaria, and Man: A History of the Hostilities Since 1880, John Murray, ISBN 0-71953-780-8, London, UK.

[58] Hauser, R., Altshul, L., Chen, Z., Ryan, L., Overstreet, J., Schiff, I. & Christiani D.C. (2002). Environmental organochlorines and semen quality: results of a pilot study. *Environmental Health Perspectives*, Vol.110, pp. 229–233, ISSN 0091-6765

[59] Hauser, R., Chen, Z., Pothier, L., Ryan, L. & Altshul, L. (2003). The relationship between human semen parameters and environmental exposure to polychlorinated biphenyls and p,pV-DDE. *Environmental Health Perspectives*, Vol.111, pp. 1505–1511, ISSN 0091-6765

[60] Hemingway, J. & Ranson, H. (2000). Insecticide resistance in insect vectors of human disease. *Annual Review of Entomology*, Vol.45, pp. 371–391, ISSN 1545-4487

[61] Hemingway, J., Beaty, B.J., Rowland, M., Scott, T.W. & Sharp, B.L. (2006). The Innovative Vector Control Consortium: improved control of mosquito-borne diseases. *Trends in Parasitology*, Vol.22, pp. 308–312, ISSN1471-4922

[62] Iscan, M., Coban, T., Cok, I., Bulbul, D., Eke, B.C. & Burgaz, S. (2002). The organochlorine pesticide residues and antioxidant enzyme activities in human breast tumors: is there any association? *Breast Cancer Research and Treatment*, Vol.72, pp. 173–182, ISSN 0167-6806

[63] Ito, J., Ghosh, A., Moreira, L.A., Wimmer, E.A. & Jacobs-Lorena, M. (2002). Transgenic anopheline mosquitoes impaired in transmission of a malaria parasite. *Nature*, Vol. 417, No.6887, pp. 452–455, ISSN 0028-0836

[64] Jensen, S., Johnels, A., Olsson, M. & Otterlind, G. (1969). DDTand PCB in marine animals from Swedish waters. *Nature*, Vol.223, No.5216, pp. 753–754, ISSN 0028-0836

[65] Keiser, J., Utzinger, J. & Singer, B.H. (2002). The potential of intermittent irrigation for increasing rice yields, lowering water consumption, reducing methane emissions, and controlling malaria in African rice fields. *Journal of the American Mosquito Control Association*, Vol.18, pp. 329–340, ISSN 1046-3607

[66] Keiser, J., Singer, B.H. & Utzinger, J. (2005). Reducing the burden of malaria in different eco-epidemiological settings with environmental management: a systematic review. *The Lancet Infectious Diseases*, Vol.5, pp.695–708, ISSN 1473-3099

[67] Kelce, W., Stone, C., Laws, S., Gray, L., Kemppainen, J. & Wilson, E. (1995). Persistent DDT metabolite, p,pV-DDE is a potent androgen receptor antagonist. *Nature*, Vol. 375, pp. 581–585, ISSN 0028-0836

[68] Khadjavi, A., Giribaldi, G. & Prato, M. (2010). From control to eradication of malaria: the end of being stuck in second gear? *Asian Pacific Journal of Tropical Medicine*, Vol.3, No.5, pp. 412-420, ISSN 1995-7645

[69] Korrick, S.A., Chen, C., Damokosh, A.I., Ni, J., Liu, X., Cho, S.I., Altshul, L., Ryan, L. & Xu, X. (2001). Association of DDT with spontaneous abortion: a case-control study. *Annals of Epidemiology*, Vol.11, pp. 491–496, ISSN 1047-2797

[70] Lacey, L.A. (2007). *Bacillus thuringiensis* serovariety *israelensis* and *Bacillus sphaericus* for mosquito control. *Journal of the American Mosquito Control Association*, Vol.23 (suppl 2), pp. 133–163, ISSN 1046-3607

[71] Langer, P., Kocan, A., Tajtakova, M., Petrik, J., Chovancova, J., Drobna, B., Jursa, S., Pavuk, M., Koska, J., Trnovec, T., Sebokova,E. & Klimes, I. (2003). Possible effects of polychlorinated biphenyls and organochlorinated pesticides on the thyroid after long-term exposure to heavy environmental pollution. *Journal of Occupational & Environmental Medicine*, Vol.45, pp. 526–532, ISSN 1536-5948

[72] Lebel, G., Dodin, S., Ayotte, P., Marcoux, S., Ferron, L.A. & Dewailly, E. (1998). Organochlorine exposure and the risk of endometriosis. *Fertility and Sterility*, Vol.69, pp. 221–228, ISSN 1556-5653

[73] Lengeler, C. (2004). Insecticide-treated bednets and curtains for preventing malaria. *Cochrane Database of Systematic Reviews*, Vol.2, pp. CD000363, ISSN 1469-493X

[74] Lewis, K. (2008). DDT stalemate stymies malaria control initiative. *Canadian Medical Association Journal*, Vol.179, No.10, pp. 999–1000, ISSN 1488-2329

[75] Lines, J. (1996). Mosquito nets and insecticides for net treatment: a discussion of existing and potential distribution systems in Africa. *Tropical Medicine & International Health*, Vol.1, No.5, pp. 616–632, ISSN 1365-3156

[76] Longnecker, M.P. (2005). Invited commentary: why DDT matters now. *The American Journal of Epidemiology*, Vol.162, pp. 726–728, ISSN 1476-6256

[77] Longnecker, M.P., Klebanoff, M.A., Brock, J.W., Zhou, H., Gray, K.A., Needham, L.L. & Wilcox, A.J. (2002). Maternal serum level of 1,1-dichloro-2,2-bis(p-chlorophenyl)ethylene and risk of cryptorchidism, hypospadias, and polythelia among male offspring. *The American Journal of Epidemiology*, Vol.155, pp. 313–322, ISSN 1476-6256

[78] Longnecker, M.P., Klebanoff, M.A., Zhou, H. & Brock, J.W. (2001). Association between maternal serum concentration of the DDT metabolite DDE and preterm and small-for-gestational-age babies at birth. *The Lancet*, Vol.358, pp. 110–114, ISSN 1474-547X

[79] MacDonald G. (1957). The Epidemiology and Control of Malaria. London, Oxford University Press.

[80] Martinez Torres, D., Chandre, F., Williamson, M.S., Darriet, F., Bergé, J.B., Devonshire, A.L., Guillet, P., Pasteur, N. & Pauron, D. (1998). Molecular characterization of pyrethroid knockdown resistance (kdr) in the major malaria vector Anopheles gambiae s.s. *Insect Molecular Biology*, Vol.7, No.2, pp. 179–184, ISSN 1365-2583

[81] Ménard, R. (2005). Medicine: knockout malaria vaccine? *Nature*, Vol.13, No.433, pp. 113-114, ISSN 0028-0836

[82] Metcalf, R.L. (1989). Insect Resistance to Insecticides. *Pesticide Science*, Vol.26, pp. 333-358, ISSN 1096-9063

[83] Munga, S., Vulule, J. & Allan, R. (2009). Evaluation of insecticide treated wall lining materials used in traditional rural African houses, Proceedings of 5th MIM Pan-African malaria conference, Nairobi, Kenya, November 2009, Poster No. 734, pp. 200

[84] Mussalo-Rauhamaa, H. (1991). Partitioning and levels of neutral organochlorine compounds in human serum, blood cells, and adipose and liver tissue. *Science of the Total Environment*, Vol.103, pp. 159–175, ISSN 0048-9697

[85] N'Guessan, R., Boko, P., Odjo, A., Akogbeto, M., Yates, A. & Rowland, M. (2007a). Chlorfenapyr: a pyrrole insecticide for the control of pyrethroid or DDT resistant *Anopheles gambiae* (Diptera: Culicidae) mosquitoes. *Acta Tropica*, Vol.102, pp. 69–78, ISSN 0001-706X

[86] N'Guessan, R., Corbel, V., Bonnet, J., Yates, A., Asidi, A., Boko, P., Odjo, A., Akogbé-to, M. & Rowland, M. (2007b). Evaluation of indoxacarb, an oxadiazine insecticide for the control of pyrethroid-resistant *Anopheles gambiae* (Diptera: Culicidae). *Journal of Medical Entomology*, Vol.44, pp. 270–276, ISSN 0022-2585

[87] Nilsen, T. & Vatten, L. (2000). A prospective study of lifestyle factors and the risk of pancreatic cancer in Nord-Trondelag, Norway. *Cancer Causes & Control*, Vol.11, pp. 645–652, ISSN 1573-7225

[88] PAN (Pesticide Action Network). (2012). Available from http://www.pan-uk.org/

[89] PANUPS (Pesticide Action Network North America) (1996). Global Distribution of organochlorines,, San Francisco, US.

[90] Pates, H. & Curtis, C. (2005). Mosquito behavior and vector control. *Annual Review of Entomology*, Vol.50, pp. 53–70, ISSN 1545-4487

[91] Peakall, D.B. (1969). Effect of DDT on calcium uptake and vitamin D metabolism in birds. *Nature*, Vol.224, pp. 1219–1220, ISSN 0028-0836

[92] Porta, M., Malats, N., Jarriod, M., Grimalt, J.O., Rifa, J., Carrato, A., Guarner, L., Salas, A., Santiago-Silva, M., Corominas, J.M., Andreu, M. & Real FX. (1999). Serum concentrations of organochlorine compounds and K-ras mutations in exocrine pancreatic cancer. *The Lancet*, Vol.354, pp. 2125–2129, ISSN 1474-547X

[93] Prato, M., Khadjavi, A., Mandili, G., Minero, V.G. & Giribaldi, G. (2012). Insecticides as Strategic Weapons for Malaria Vector Control, In: *Insecticides - Advances in Integrated Pest Management*, Perveen, F. (Ed.), pp.91-114, InTech, ISBN: 978-953-307-780-2, Rijeka, Croatia, Available from http://www.intechopen.com/books/insecticides-advances-in-integrated-pest-management/insecticides-as-strategic-weapons-for-malaria-vector-control

[94] Ramade, F. (1987). Ecotoxicology. New York, John Wiley & Sons.

[95] Ratcliffe, D. (1970). Changes attributable to pesticides in egg breakage frequency and eggshell thickness in some British birds. *Journal of Applied Ecology*, Vol.7, pp. 67–115, ISSN 0021-8790

[96] Rathore, M., Bhatnagar, P., Mathur, D. & Saxena, G.N. (2002). Burden of organochlorine pesticides in blood and its effect on thyroid hormones in women. *Science of the Total Environment*, Vol.295, pp. 207–215, ISSN 0048-9697

[97] Ritter, L., Solomon, K.R., Forget, J., Stemeroff, M. & O'Leary, C. (1995). Persistent Organic Pollutants. Geneva: International Programme on Chemical Safety

[98] Roberts, D., Laughlin, L.L., Hsheih, P. & Legters, L.J. (1997). DDT, global strategies, and a malaria control crisis in South America. *Emerging Infectious Diseases*, Vol.3, No. 3, pp.295–302, ISSN 1080-6059

[99] Roberts, L. & Enserink, M. (2007). Did they really say… eradication? *Science*, Vol.318, No.5856, pp. 1544-1545, ISSN 1095-9203

[100] Rogier, C. & Hommel, M. (March 2011). *Plasmodium* life-cycle and natural history of malaria, In: Impact malaria sanofi Aventis' commitment, 9 March 2011, Available from   http://www.impact-malaria.com/iml/cx/en/layout.jsp?cnt=2FAECA4C-CA72-4C97-9090-B725867E1579

[101] Sadasivaiah, S., Tozan, Y. & Breman, J.G. (2007). Dichlorodiphenyltrichloroethane (DDT) for Indoor Residual Spraying in Africa: How Can It Be Used for Malaria Control? *American Journal of Tropical Medicineand Hygiene*, Vol.77, No.6, pp. 249–263, ISSN 0002-9637

[102] SADC (Southern African Development Community) (2011). Letter from SADC to A Steiner, Executive Director, United Nations Environment Programme, Notification of SADC Countries Respecting DDT Use and Production, 5 April 2011, Available from http://www.fightingmalaria.org/pdfs/sadclettertounep.pdf

[103] Saxena, M., Siddiqui, M., Bhargava, A., Seth, T., Krishnamurti, C. & Kutty, D. (1980). Role of chlorinated hydrocarbon pesticides in abortions and premature labour. *Toxicology*, Vol.17, pp. 323–331, ISSN 0300-483X

[104] Schellenberg, J.R.M., Abdulla, S., Nathan, R., Mukasa, O., Marchant, T.J., Kikumbih, N., Mushi, A.K., Mponda, H., Minja, H., Mshinda, H., Tanner, M. & Lengeler, C. (2001). Effect of large-scale social marketing of insecticide-treated nets on child survival in rural Tanzania. *The Lancet*, Vol.340, pp.1241–1247, ISSN 0140-6736

[105] Scholte, E.J., Ng'habi, K., Kihonda, J., Takken, W., Paaijmans, K., Abdulla, S., Kelleen, G.F. & Knols, B.G. (2005). An entomopathogenic fungus for control of adult African malaria mosquitoes. *Science*. Vol.308, pp. 1641–1642, ISSN 0036-8075

[106] Settimi, L., Masina, A., Andrion, A. & Axelson, O. (2003). Prostate cancer and exposure to pesticides in agricultural settings. *International Journal of Cancer*, Vol.104, pp. 458–461, ISSN 1097-0215

[107] Sever, L.E., Arbuckle, T.E. & Sweeney, A. (1997). Reproductive and developmental effects of occupational pesticide exposure: the epidemiologic evidence. *Occupational Medicine*, Vol.12, pp. 305–325, ISSN 1471-8405

[108] Sharpe, R. & Skakkebaek, N. (1993). Are oestrogens involved in falling sperm counts and disorders of the male reproductive tract? *The Lancet*, Vol.341, pp. 1392–1395, ISSN 1474-547X

[109] Siddiqui, M.K., Nigam, U., Srivastava, S., Tejeshwar, D.S. & Chandrawati R. (2002). Association of maternal blood pressure and hemoglobin level with organochlorines in human milk. *Human & Experimental Toxicology*, Vol.21, No.1, pp. 1–6, ISSN 1477-0903

[110] Siddiqui, M.K., Srivastava, S., Srivastava, S.P., Mehrotra, P.K., Mathur, N. & Tandon, I. (2003). Persistent chlorinated pesticides and intra-uterine foetal growth retardation: a possible association. *International Archives of Occupational and Environmental Health*, Vol.76, pp. 75–80, ISSN 1432-1246

[111] Snedeker, S.M. (2001). Pesticides and breast cancer risk: a review of DDT, DDE, and dieldrin. *Environmental Health Perspectives*, Vol.1, pp. 35–47, ISSN 0091-6765

[112] Sturgeon, S.R., Brock, J.W., Potischman, N., Needham, L.L., Rothman, N., Brinton, L.A. & Hoover, R.N. (1998). Serum concentrations of organochlorine compounds and endometrial cancer risk (United States). *Cancer Causes & Control*, Vol.9, pp. 417–424, ISSN 1573-7225

[113] Tren, R. & Roberts, D. (2010). DDT and malaria prevention. *Environmental Health Perspectives*, Vol.118, No.1, pp.A 14–15, author reply pp. A15-16, ISSN 0091-6765

[114] Tren, R. & Roberts, D. (2011). DDT paradox. *Environmental Health Perspectives*, Vol. 119, No.10, pp. A423-424, author reply pp. A424-425, ISSN 0091-6765

[115] Trieu, A., Kayala, M.A., Burk, C., Molina, D.M., Freilich, D.A., Richie, T.L., Baldi, P., Felgner, P.L. & Doolan, D.L. (2011) Sterile protective immunity to malaria is associated with a panel of novel P.falciparum antigens. *Molecular and Cellular Proteomics*, Vol. 10, No.9, pp.M111.00794, ISSN 1535-9484

[116] Turusov, V., Rakitsky, V. & Tomatis, L. (2002). Dichlorodiphenyltrichloroethane (DDT): ubiquity, persistence, and risks. *Environmental Health Perspectives*, Vol.110, No.2, pp. 125–128, ISSN 1552-9924

[117] United Nations Environment Programme (UNEP). (1997). Governing Council, 19th session, see also UNEP GC decisions 18/32, 18/31

[118] United Nations Environment Programme (UNEP). (1997). Status Report on UNEP's and other related activities on POPs

[119] United Nations Environment Programme (2007). Future Plans for Work on DDT Elimination. A Stockholm Convention Secretariat Position Paper, November 2007. Geneva: Secretariat of the Stockholm Convention, Available from http://chm.pops.int/Portals/0/Repository/DDT-general/UNEP-POPS-DDT-PROP-SSCPP.English.PDF

[120] Utzinger, J., Tozan, Y. & Singer, B.H. (2001). Efficacy and cost-effectiveness of environmental management for malaria control. *Tropical Medicine & International Health*, Vol.6, pp. 677–687, ISSN 1365-3156

[121] van den Berg H. (2009). Global status of DDT and its alternatives for use in vector control to prevent disease. *Environmental Health Perspectives*, Vol.117, No.11, pp. 1656–1663, ISSN 0091-6765

[122] van Wendel de Joode, B., Wesseling, C., Kromhout, H., Monge, P., Garcia, M. & Mergler, D. (2001). Chronic nervous-system effects of longterm occupational exposure to DDT. *The Lancet*, Vol.357, pp. 1014–1016, ISSN 1474-547X

[123] Varca, L.M. & Magallona, E.D. (1994). Dissipation and degradation of DDT and DDE in Philippine soil under field conditions. *Journal of Environmental Science and Health*, Vol.29, pp. 25–35, ISSN 0360-1234

[124] Walker, K. (2000). Cost-comparison of DDT and alternative insecticides for malaria control. *Medical and Veterinary Entomology* Vol.14, No.4, pp. 345-354, ISSN 0269-283X

[125] Walker, K. & Lynch, M. (2007). Contributions of *Anopheles* larval control to malaria suppression in tropical Africa: review of achievements and potential. *Medical and Veterinary Entomology*, Vol.21, pp.2–21, ISSN 1365-2915

[126] Wandiga, S.O. (2001). Use and distribution of organochlorine pesticides. The future in Africa. *Pure and Applied Chemistry*, Vol.73, pp. 1147–1155, ISSN 1365-3075

[127] Weiderpass, E., Adami, H.O., Baron, J.A., Wicklund-Glynn, A., Aune, M., Atuma, S. & Persson, I. (2000). Organochlorines and endometrial cancer risk. *Cancer Epidemiology, Biomarkers & Prevention*, Vol.9, pp. 487–493, ISSN 1538-7755

[128] WHO (World health Organization) (1996). The WHO Recommended Classification of Pesticides by Hazard and Guidelines to Classification 1996-1997, UNEP, ILO, WHO, Geneva, Switzerland

[129] WHO (2000). Manual for indoor residual spraying. Geneva:World Health Organization.

[130] WHO (2005). Regional strategic framework for scaling up the use of insecticide-treated nets, In: *Insecticides treated materials*, n.d., Available from http://www.searo.who.int/LinkFiles/ Reports_MAL-239_&_VBC-87.pdf.

[131] WHO (2006a). WHO gives indoor use of DDT a clean bill of health for controlling malaria, In: *World Health Organization*, (20 September 2006), Available from http://www.who.int/mediacentre/ news/releases/2006/pr50/en/print.html

[132] WHO (2006b). Indoor residual spraying, use of indoor residual spraying for scaling up global malaria control and elimination, In: *Indoor residual spraying*, World Health Organization, (11 December 2006), Available from http://www.who.int/malaria/vector_control/irs/en/

[133] WHO (2011a). World malaria report 2011 Available from http://www.who.int/malaria/world_malaria_report_2011/en

[134] WHO (2011b) DDT in Indoor Residual Spraying: Human Health Aspects. Environmental Health Criteria 241. Geneva:WHO. Available from http://www.who.int/ipcs/publications/ehc/ehc241.pdf

[135] Yanez, L., Ortiz-Perez, D., Batres, L.E., Borja-Aburto, V.H. & Diaz-Barriga, F. (2002). Levels of dichlorodiphenyltrichloroethane and deltamethrin in humans and environmental samples in malarious areas of Mexico. *Environmental Research*, Vol.88, pp. 174–181, ISSN 0013-9351

[136] Zwiebel, L.J. & Takken, W. (2004). Olfactory regulation of mosquito-host interactions. *Insect Biochemistry and Molecular Biology*, Vol.34, pp.645–652, ISSN 0965-1748

Chapter 2

Insecticide Residuality of
Mexican Populations Occupationally Exposed

María-Lourdes Aldana-Madrid,
María-Isabel Silveira-Gramont,
Fabiola-Gabriela Zuno-Floriano and
Guillermo Rodríguez-Olibarría

Additional information is available at the end of the chapter

1. Introduction

In this chapter the information generated by two research projects conducted in Sonora, Mexico (2003-2010) with the objective to assess the presence of insecticide residues (organo-chlorine and organophosphates) in corporal fluids of two population groups that live and work in agricultural areas (agricultural workers) are described. The presence of residues in body fluids (blood, urine, semen and breast milk) will be related to social, labor, environmental factors, health status (obtained from surveys, and clinical histories), and biochemical and biological indicators, with the purpose of elucidating the degree and persistence of the exposure to these insecticides.

In the countries in development as Mexico, the handling of toxic compounds as pesticides is inadequate; it is possible that people can be exposed to higher concentrations than that allowed by the maximum limits (LMP), as they demonstrate it studies carried out with children in San Luis Potosí state [1], in some endemic areas of malaria in Mexico like Quintana Roo [2], Chiapas and Oaxaca [3-5], and in labor exposed people of Sonora state [6,7]. This evidence suggests that the populations that work in agricultural fields, as well as those that inhabit the surrounding area could have higher exposure risk, as well as chronic contamination that the populations with a basal exposure.

The state of Sonora is amongst the regions of Mexico with more pesticide use; it is calculated that 80% of the total applied in the country is for the production of grains and export vegetables [8]. There are not reliable statistics of pesticide intoxications in rural areas, and there are not epidemiologic studies to detect chronic effects of the pesticides; those should exist at least for the agricultural journeymen and for vulnerable groups, since they lack elementary protection, and don't have the correct information about pesticide toxicity.

The exposure doses can be small but persistent, causing chronic health problems [8]. DDT (bis[4-chlorophenyl]-1,1,1-trichloroethane, also called dichlorodiphenyl trichloro-ethane); was first used to protect military areas and personnel against malaria, typhus, and other vector-terminal diseases [9]. In Mesoamerica (Mexico, Costa Rica, El Salvador, Guatemala, Honduras, Nicaragua and Panama) DDT was used until the year 2000; Mexico and Nicaragua being the last nations that applied the insecticide in agriculture and for the control of malaria (the amount used for Mexico is approximately 69,545 tons between 1957 and 2000) [10].

Technical-grade DDT contains 65-80% p,p'-DDT, 15-21% o,p'-DDT, and up to 4% p,p'-DDD (bis[4-chlorophenyl]-1,1,-dichloroethane)[11]. When sprayed, DDT can drift, sometimes for long distances. In the soil, the compound can evaporate or attach to wind-blown dust. In the environment, DDT breaks down to p,p'-DDE (bis[4-chlorophenyl]-1,1-dichloroethene) [12], an extremely stable compound that resists further environmental breakdown or metabolism by organism; DDE is the form usually found in human tissue in the highest concentration, especially in areas where there has been no recent use of the parent compound [9]. DDT and DDE can also be transferred from the placenta and breast milk to fetuses and infants. Although some ingested DDT is converted to DDA (bis[4-chlorophenyl]-acetic acid) and excreted, any non-metabolized DDT and any DDE produced is stored in fat, as is all absorbed DDE, which cannot be metabolized. DDT and DDE are highly soluble in lipid; their concentrations are much higher in human adipose tissues (about 65% fat) than in breast milk (2.5-4% fat), and higher in breast milk than in blood or serum (1% fat) [13]. DDT and DDE concentrations increase with age [14].

The use of DDT in Central and South America, Mexico, Africa, and some Asian countries where this has been used for vector control in the past 5-10 years, DDT concentrations in human tissues remain high. For example, in Mexico, the total DDT concentration in breast milk fat was 5.7 µg/g in 1994-95 and 4.7 µg/g in 1997-98 [15]. Others Mexican data where workers used DDT to control mosquitoes, have very high DDT concentrations. Mexican data revealed that the geometric mean of total DDT was 104.48 µg/g in adipose tissue of 40 DDT sprayers in 1996 [16]; however in Finland, USA, and Canada, the value was less than 1 µg/g in adipose tissue in the general population [17]. In another Mexican study, the serum concentration of p,p'-DDE was much higher in DDT sprayers (188 µg/L) than in children (87 µg/L); also in adults (61 µg/L) who lived in sprayed houses, but were not otherwise exposed to DDT [3].

The organophosphate insecticides have the advantage of low environmental persistence over the chlorinated pesticide compounds. However, studies carried out in individuals with exposure to insecticides in Mexico, and other countries, associate the pesticide exposure with adverse health effects, as much in the humans as in experimental animals. These damages can be evident by the presence of certain biochemical indicators in the different biological fluids, and for the detection of morphological, histological, and molecular changes in target organs [2-4,6,7].

The degree of pesticide contamination depends on several factors, such as the formulation of the pesticide, the active ingredient, the time of exposure, the direct or indirect contact, the quantity used, the pesticides mixtures, the climate and season of the year when it's applied, and the person's age, amongst others [18, 19]. There are environmental indicators, health indicators, and other elements that help determine the exposure risk, such as the person residence and occupational history, the clinical history, as well as the presence of the pesticides studied in drinking water, in the ground, in the atmosphere, and in the fresh or processed foods in the region where the studied populations inhabit. The exposure can be increased with the daily time dedicated to the activity, as well as the years of work, the exposure form, the use of protective gear, and/or the physical proximity of the housing to agricultural fields [20-22].

Due to the previously mentioned situations it was considered important to study the degree of exposure to pesticides on workers in those agricultural areas that also reside in their proximity.

2. Materials and methods

2.1. Site description

Sonora State territory has 179,355 km^2 and it is located in Sierra Madre Occidental; geographically it is north 32° 29', to south 26° 18' of north latitude, to east 108° 25', to west 115° 03'. The weather in coastal of Sonora is dry. The average annual temperature is around 22°C, being the average maximum 38°C (June and July) and the average minimum temperature is 5°C (January). Only 7% of the land is appropriate for agricultural use and ninety five percent of this area is irrigated. In July and August the rain reaches 450 mm. The weather in Sonora State restricts the agricultural activities. However, in villages Yaqui and Mayo, valleys of Hermosillo, Caborca and Guaymas the major crops with irrigation are wheat, cotton, safflower, watermelon, sesame, garbanzo, sorghum, corn and vine [23]. The agriculture in the south of Sonora is based on 90% of crops such as corn, wheat, oleaginous and cotton [24].

2.2. Population study

Group 1, field workers. The municipal headboards included in this group are localized in the following coordinates: Obregon city 27° 29' north latitude and 109°39' west latitude with a height of 10 m above sea level and Navojoa 27° 05' north latitude and 109°39' west latitude with 40 m above sea level [23].

The town council Cajeme has a population of 175,177 men, from this population, 6,983 live in Yaqui town. In Huatabampo there are 38,563 men, specifically in "Jupare" (1,026 men). Navojoa has a population of 69,341 from this population, 445 men live in 5 de Ju-

nio cooperative [25]. In this study participated 37 men from Yaqui town, 19 from "Jupare" and 21 from "5 de Junio" cooperative, this is 0.53%, 1.85% and 4.7% of total
population of men, respectively.

2.3. Group 2, nursing mothers

The women that participated in this study were located at Pesqueira community. The community is located between 28 and 30º parallel north latitude in Sonora, Mexico [23]. The
weather in this region is dry with rain in summer [23]. The women are dedicated to cultivat-
ing and packing table grape. When this study was conducted the population in Pesqueira
was 3,648 residents; 47.8% women, from this percentage only 10% were in reproductive age
[25]. In the Health center there was a record of 20% women included in the breastfeeding
program. However, 26% women in reproductive age (exposed or working with pesticides)
participated in the present study, being 1.4% of total population. It is probably that permanent residents of agricultural areas are chronically exposed to chemical residues through
wind, drinking water and even clothing from field workers.

2.4. Participation and surveys

Men and women (nursing period) that voluntarily participated in this study, signed a format according to the norms of Mexican Secretary of Health (SS). Participants filled up a survey; they provided demographic data (age, marital status, residence, residence time) and
also information related to work history, pesticides exposure, clinical history, issues related
to sexuality, pathology and drugs addiction (alcohol, tobacco, cocaine and marihuana,
among others). People with drugs addictions or health issues that could have influence with
the biochemical determinations of the body fluids were excluded from this research.

2.5. Blood, urine and semen sampling

Blood, urine and semen samples were collected from members of group 1. For hematic biometry analysis, blood chemistry and biochemical indicators, samples were taken with empty
stomach. Blood samples were taken by venous puncture and collected in to vacuum tubes
(Vacutainer^{MR}) with and without anticoagulant. Once blood was coagulated, it was centrifuged and supernant was transferred in to a new tube for analysis.

Urine samples were collected in a sterile container and kept at 4 °C until analysis. Semen samples were collected in sterile container (including a code, date and time when
were collected) by the participant at home. Samples were analyzed no more than 24 h after sampling.

2.5.1. Breast milk sampling

Breast milk was collected either manually or with a breast milk collector in to a 50 mL coni-
cal glass tubes (wrapped with aluminum foil). Samples were kept at -20ºC until analysis.

2.5.2. Blood analysis

Blood samples were analyzed in laboratory of General Hospital SS in Cd. Obregon and the laboratory of General Hospital of Navojoa. Blood analysis included hematic biometry using an analyzer Sysmex K-4500, blood chemistry test (glucose, urea, creatinine, uric acid, cholesterol and triglycerides) and total proteins (albumin and globulin) using an analyzer HITACHI 911 and also determination of enzymes in serum such as levels of plasma cholinesterase (Randox[MR]), alkaline phosphatase (Roche[MR]), transaminases, and superoxide dismutase (SOD, Randox[MR]).

2.5.3. Urine analysis

Urine was analyzed by two types of analysis; biochemical and microscopic analysis. Biochemical analysis included glucose, proteins, bilirubin, kenotic bodies, urobilinogen (combo test-10). The macroscopic analysis included number of bacteria, erythrocytes, leucocytes, crystals, epithelial cells, etc.

2.5.4. Semen analysis

Analysis of semen was carried out by using international standardized techniques [26]. Macroscopic and microscopic analyses were included; in the first one liquefaction, aspect, viscosity, pH and volume were determined. In the microscopic analysis motility, viability, presence of leucocytes, erythrocytes, germinal cells, dendrites, agglutination (specific and unspecific), number of spermatozoa and morphology were determined.

2.5.5. Insecticides extraction

Blood, urine and semen. Samples were analyzed following the methodology proposed in [27]. Briefly, 0.5 mL of sample were taken and added 4 mL hexane; the mixture was shaken for 15 seconds and then centrifuged for 2.30 minutes at 2500 rpm. The supernant was transferred to a tube and added 1 mL of 5% K_2CO_3 and 4 mL hexane. Centrifuged for 2.30 minutes at 2500 rpm, supernant was transferred to a tube and evaporated to dryness. Extract was dissolved with 100 µL hexane and analyzed by gas chromatography. *Breast milk* was analyzed using a matrix solid-phase dispersion technique [28].

2.6. Insecticides residues analyses in body fluids

The analytical standards were from Chem Service (West Chester, PA). Quantitation of insecticides was by comparison of five –point calibration curve. The detection ranges used for calibration curves were 50-0.1 µg/L. The average percent recoveries for organochlorine pesticides were p,p'-DDD 95%, p,p'-DDE 98% and p,p'DDT 105%, for organophosphates were diazinon 99%, clorpyrifos 91%, malathion 106% and parathion 92%.

Quantitative analyses were performed by gas chromatography (GC) using a Varian model CP-3800 equipped with an electron capture detector (ECD)(USA). The insecticides were separated using VA-1701 (Varian, 30 m x 0.25 mm) capillary column. The injection volume was

1 µL. Nitrogen (purity 99.999 %) was used as the carrier gas at a flow of 1.5 mL/min. The injector temperature was 180°C and the detector temperature was 300°C. The temperature program was as follows: initial temperature 220°C, increasing temperature at 9°C min^{-1} until the final temperature of 300°C was reached. Data was analyzed using a program Star Chromatography Workstation 5.51.

2.7. Design and analysis of the studies

The design of the study 1 was of the type "Case/Control", where 77 men integrated case group and 17 the control group. Participants of both groups were the same ages (18 to 70 years old). Control group did not have evidence of pesticides exposure. The sample size for the cases group represented approximately 5% of the total male population's in the range of ages selected in the study.

The study 2 was integrated by 39 nursing mothers between 17 and 39 years old, selected randomly among those that accepted to participate. All the other characteristics were similar to study 1.

The nominal data were analyzed by group (in study 1) using contingency tables and Chi-square statistics. Continuous numeric data (age, height, weight, etc.) were reported with descriptive statistics (minimum, maximum, mean, median, and standard deviation). In study 1 there were analysis of variance comparing groups (cases vs controls) for the numeric and nominal variables. Several exposure indicators (reported by the literature, erythrocytes, VCM, and RDW) were analyzed by linear regression versus time of exposure, and pesticide amount on a particular body fluid. Also, multiple correlation coefficients were estimated between several biological indicators.

For the study 2 beside of the descriptive statistics, some relationships were evaluated such as the use of protective gear, age, number of years of exposure, children's number, among others. The pesticide residues in breast milk were compared with the maximum residual limits stated by international organizations.

3. Results and discussion

Based on the data obtained in the present study, it was observed that the case group had similar demographic characteristics to the control group but the last one without any pesticide exposure.

3.1. Description of the population based on the surveys and clinical history

Group 1. Field workers. 94 personal surveys were made from which 68% provided data related to medical history. Only 71% of the participants provided blood samples, 69% urine samples and 46% seminal fluid samples. A small number of medical histories and samples were obtained since the participants had the liberty to leave the study at any time.

In the case group, a total of 77 men participated; the majority maintained contact with organophosphate pesticides; the average age of this group was 40 years and 11 years of residency in the site of the study with a maximum of 45 years. The average work years with pesticide exposure was 28, with a maximum of 50 years working in agriculture. Based on data from work history a 62% of the cases had contact with pesticides; from this percentage 43% applied them, 27% works in the places where they were applied and 30% works where they are applied. Only 18% of the field workers uses protection while applying pesticides (like gloves, special clothing for welding fumes, paint fumes o foundry fumes). These results suggest that field workers are chronically exposed to pesticides due to few safety precautions are taken to handle them. There for, it is of importance that the field workers receive training to be aware of the possible health issues related to pesticides exposure.

In the case group besides being in contact with pesticides during work activities they are also in their place of residency; considering this background, the time of exposure is 16 years in average and 65% of them apply insecticides in their homes; 27% are applied with an annual frequency, 23% are applied semiannually, while 10% every 3 months and 8% monthly.

Some factors can exacerbate the toxicological effects caused by pesticide exposure, such as the consumption of alcohol and drugs. The present study found that 69% of the individuals of the case group consume alcohol with a monthly frequency, 16% consume less than 5 cigarettes daily, 8% and 6% less than 20 cigarettes and the rest only 1 cigarette, 4% consumes cocaine, none of the cases consumes marihuana, nor intravenous drugs. Erection and ejaculation problems (4%) were found in case group. Case and control groups had problems having children (6%). Unlike the control group, the case group presented sexual transmission diseases; around 4% had gonorrhea. Both groups have children with congenital health problems (approximately 6%).

Some of the reported symptoms in the case group were cramps (61%), tiredness and weakness (53%), blurred vision (45%), sweating (45%), tearing (43%), nervousness (38%), dizziness (37%) and tingling in the extremities (37%). According to literature it can be considered that pesticide intoxication is nonspecific and produce the subclinical symptoms identified in the present study in addition to anorexia, insomnia, digestive alterations and itching of skin and mucous [11]. Mostly the symptoms caused by pesticides exposure are diagnosed as common cold or flu [29]. This symptomatology is not produced at the same time because every chemical product acts in a different way and will differ in each of the persons with a chronic exposure. In the present study, during the physical auscultation, the average weight and height in the case group and control group was 82 and 81 kg and 1.72 m, respectively. The vital signs were normal for both groups; 70 and 80 pulsations/min (normal value 70-80 pulsations/min), respiratory rate 20 breaths/min (normal 12-20 breaths/min) and blood pressure 120/180 (normal 120/180).

Group 2. Nursing women, description based on surveys. A total of 51 surveys were made to nursing women, 79% of them have been living in this community for more than five years. The average age was 24 years while the median was 23. The highest age was 39 and the lowest 16. The average body weight of women was 82 kg (±15.9), with an average height of 1.72 m (±0.07). The average number of children was three, 98% were married (including the ones that live in free union) and only 2% were single (this includes also the ones that are divorced). The 72% of the women were housewives, but 92% of them mentioned to have worked in agriculture. The 77% of the participants were in contact with pesticides; 69% applied in more than one occasion and only 34% used protective clothing while applying (gloves, special clothing for their work and mask).

The 53% applied pesticides in their home; 38% with an annual application, 43% monthly and 19% weekly. Insecticides applied at home were pyrethroids (20%), and organophosphates (14%).

According to the literature, intoxication by pesticides is nonspecific and produce symptoms like: excitability, tremors, sweating, tiredness, dizziness, headache and convulsions; in women they can also cause a decrease in the duration of breastfeeding [11]. The symptoms present in the participants of this study were fatigue (70%), headache (62%) and perspiration (46%). Around 76% of women (39 of 51) agreed to donate breast milk, 14% (7 of 51) of the women decided to retire and not to collaborate more in the research and the remaining 10% (5 of 51), were not producing the necessary amount for analysis.

3.2. Blood analysis

Significant differences were observed between both groups (case and control) regarding the number of erythrocytes, mean corpuscular hemoglobin concentration (MCHC), and red blood distribution width (RDW); these values were lower for the case group (Table 1). Regarding the obtained results in the chemical blood analysis, both groups presented levels within the normal values for the measured biochemical indicators. However, it is important to mention that statistical differences were observed between the groups with respect to the concentration of total protein, albumin, alkaline phosphatase, glutamic oxaloacetic transaminase (SGOT or AST) and glutamic pyruvic transaminase (SGPT or ALT); the found levels in the case group were superior to control. In a study performed with pesticide factory workers exposed to carbamates, organophosphates and organochlorines superior values were observed in total proteins [30]. However, in studies conducted with experimental animals, total protein values were not altered by the presence of pyrethroid such as cypermethrin, but the albumin levels decreased at the fifth day of intoxication [31]. Some studies performed to determine the influence of pesticide residue on biochemical indicators have reported that glucose levels increased after expose experimental animals to malathion (20 μg/mL) and after few hours levels went back to normal [32]. On the other hand, it has been reported that cholesterol and triglycerides levels are inhibited after applying a daily dose of cypermethrin (a pyrethroid insecticide) to Wistar rats [31]. A study referring to the toxicological effect of polychlorinated biphenyls (PCB's) in fish, reported that PCB's caused lipid peroxidation, increased cholesterol levels in serum and in some species caused hepatic toxicity and hypertension [33]. Recent studies about the indiscriminate use of pesticides in Tasmania, Australia, have reported effects in the health of its habitants (obesity, hypertension and high cholesterol levels) [34]. Researchers have confirmed that acetylcholinesterase is an indicator of damage by organophosphate and carbamate pesticides [35], in this study the case group mentioned having contact with this substances and the levels of cholinesterase were below normal values. In previous research it was observed that chronic exposure to organophosphate insecticides is related to an increase of catalase, superoxide dismutase and glutathione peroxidase [36]. A study performed in the South India, related to the effect of pesticides on SOD, observed an increase in the levels of this enzyme parallel to the severity of the poisoning with organophosphates [37].

Analyses	Case group	Control group	Normal levels*
Leucocytes (cells/μL)	6,966.20	6,664.71	5,000 – 10,000
Erythrocytes (millions of cells/μL)[1]	4.96	5.13	4.6 - 6.2
Hemoglobin (g/dL)	14.83	15.24	13.5 - 18
Mean corpuscular hemoglobin concentration (MCHC) (%)[2]	33.06	33.52	32 - 36
RDW (fL)[3]	44.39	44.62	35 – 55

[1]p<0.06, [2]p<0.04, [3]p<0.009.

* [50].

Table 1. Results of hematic biometry analyses conducted on men exposed to pesticides from Sonora, Mexico

Analyses	Case group	Control group	Normal levels
Serum glucose (mg/dL)	105.38	102	55 - 115
Serum urea (mg/dL)	27	27	10 - 50
Creatinin (mg/dL)	0.86	0.86	0.7 - 1.2
Cholesterol (mg/dL)	195.40	180.05	< 200
Triglycerides (mg/dL)	137	131.29	< 150
Total proteins (g/dL)[1]	7.95	7.5	6.4 - 8.3
Albumin (g/dL)[2]	4.87	4.51	3.5 - 5
Globulin (g/dL)	3.09	2.98	2.3 - 3.5
Relation Albumin/Globulin	1.69	1.55	2.5
Total bilirrubin (mg/dL)[3]	0.67	0.93	< 1.1
Alkaline phosphatase (U/L)[4]	106.20	80.31	40 - 129
Cholinesterase (U/L)[5]	968-3940	3,382-8,108	4,300-10,500
Dismutase superoxide (U/mL)	273.38	275.08	164 - 240

[1]p<0.03, [2]p<0.0002, [3]p<0.039, [4]p<0.0013, [5]p<0.026.

Table 2. Levels of blood chemistry test conducted on men exposed to pesticides from Sonora, Mexico

3.3. Urine analysis

The results of the testing performed on urine for both groups were very similar and were within the normal values, it can be indicate that no abnormalities were observed in the corporal fluid that can be attributed to pesticide exposure. Besides, no statistical differences were observed amongst the study group (Table 3).

Analyses	Case group	Control group
Leucocytes *(cells/field)*	2	2
Uric acid crystals	53.33% (poor)	25% (moderate)
Calcium oxalate crystals	20% (abundant) 26.67% (moderate)	25% (abundant) 25% (moderate)
Amorphous salts	50% (abundant) 37.5% (poor) 12.5% (moderate)	57.14% (abundant) 28.57% (poor) 14.29% (moderate)
Epithelial cells	95% (poor) 5% (moderate)	87.5% (poor) 12.5% (moderate)
Bacteria	80.77% (poor)	54.55% (poor)
Mucine	59% (poor)	75% (moderate)

Table 3. Results of microscopic analyses of urine conducted in men exposed to pesticides from Sonora, Mexico

3.4. Semen analysis

Mostly all the differences between groups (case and control) were observed in the semen analyses. In table 4, it can be observed that the case group presented a lower volume, pH, sperm viability, fast and slow progressive motility and abnormalities in sperm morphology (spermatozoa macrocephalia, microcephalia, pyriform, band-like, pin-shaped, double head, tail coiled cytoplasmic droplets and amorphous). Additionally, this same group had a higher viscosity and immobility of spermatozoa. In previous studies [6], it was observed that liquefaction was affected by 32% in insecticide applicators, while the present study did not show abnormalities or significant differences between groups. In this study we observed that in the controls there were more live sperm (one third more than the majority of the cases). The percentage of the abnormalities detected in the present study was superior represented by a 35% that the one found in the control group, comparing this result with the study regarding the insecticide applicators in Hermosillo, Sonora, there was a similar behavior [6].

Analyses	Case group	Control group	Normal levels*
Liquefaction	6.9% (normal)	Normal	
Volume (mL)[1]	2.48	3.73	"/> 2
pH[2]	8.05	8.35	7.2 – 8.0
Viscosity (Filament bigger than 2 cm)	44.83% (normal)	28.57% (normal)	
Aspect	96.55% (normal) 3.45% (anormal, yellow color)	100% (normal)	Opalescent gray color (normal)
Sperm viability	45%	75%	
Sperm fast progressive motility	42.5%	52.80%	
Sperm slow progressive motility	16.88%	21.07%	
Sperm mobility	57.13%	28.93%	
Bacteria	12%	7%	
Germinal cells (Dentritus)	30.30%	7.14%	
Specific sperm agglutination espermatozoides	24.24%	-	
Normal sperm morphology	41.08%	75.92%	

[1]$p<0.03$, [2]$p<0.01$, [3]$p<0.0201$, [4]$p<0.0305$.

* [26].

Table 4. Results of semen analysis conducted on men exposed to pesticides from Sonora, Mexico

Biological fluid	Diazinon	Chlorpyrifos	Malathion	Parathion
Semen[1]	17.6	32.3	53	44
Blood[2]	0	9.1	0	9.1
Urine[3]	0	9.1	5.5	20

[1]n=11, [2]n=33, [3]n=55

Table 5. Organophosphate pesticide residues in biological fluids of field workers from Sonora, Mexico

Biological fluid	p,p'-DDT	p,p'-DDE	p,p'-DDD
Semen[1]	27.3	36.4	36.4
Blood[2]	0	15.2	3
Urine[3]	5.5	9.1	14.5

[1]n=11, [2]n=33, [3]n=55.

Table 6. Organochlorine pesticide residues in biological fluids of field workers from Sonora, Mexico

3.5. Association of exposure indicators

In the regression analysis a relationship was observed between biochemical indicators and pesticide exposure time. The biochemical indicators involved were erythrocytes, mean corpuscular volume, red blood distribution width and urea. For every year of pesticide exposure there was a decrement of 0.082 million of erythrocytes and 0.088 fL of VCM. For every year of pesticide exposure there was an increment of 0.134 fL RDW and 0.163 mg/ dL of urea. In a research performed in Spain with workers chronically exposed to pesticides, the affected biochemical indicators were urea, TGO enzymes and lactate dehydrogenase (LDH) [38]. In the figure 1, there are the associations studied between biochemical indicators and pesticides exposure time. It is important to mention that the determination of cholinesterase in this study was carried out using butyrylcholinesterase or plasma cholinesterase. In previous studies cholinesterase was associated with other biochemical indicators. The results showed that it bound to cholesterol, triglycerides and others as transaminases [38]. This was not possible to observe in the present study, due to the difference on diet and alcohol consumption.

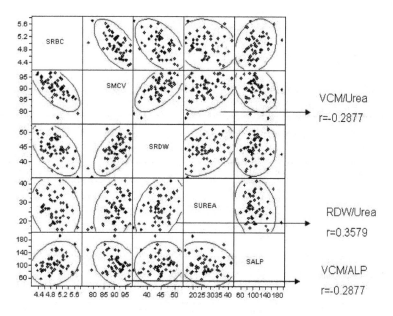

Figure 1. Correlations of blood analyses for case group

3.6. Determination of insecticides and corporal fluids

Blood, urine and semen. A total of 103 samples were analyzed: 73 (71%) were positive to some pesticide. Around 44 (60%) had organochlorine insecticides residues and 29 (40%) had organophosphorus insecticides residues. The organochlorine insecticides detected were 41%

p,p'-DDE, 39% p,p'-DDD and 20% p,p'-DDT. Regarding the organophosphate insecticides analyzed 52% had parathion, 28% chlorpyrifos, 14% malathion and 7% parathion.The highest concentration found was 7.1 µg/L of p,p'-DDE in serum and 6.4 µg/L of p,p'-DDD in urine. The highest concentration found in semen was 2.3 µg/L of p,p'-DDE. Regarding the organophosphate insecticides (chlorpyrifos, malathion and parathion) in the field workers urine the highest concentration were 3.4, 2.2 y 2.0 µg/L, respectively. This levels were considered lower in relation to other studies [39].

Breast milk. There was not detected DDT and DDT metabolites in 85.6% of breast milk samples. Although, 15.4% of samples had p,p'-DDT, p,p'-DDD and p,p'-DDE residues. The most persistent metabolite was p,p'-DDE due to its stability amongst the DDT metabolites [40-42]. The highest level found in breast milk was 9.0 µg/kg (p,p'-DDE) and the lowest level was 0.1 µg/kg (p,p'-DDT). It is important to mention that the infants fed with this contaminated breast milk were in the range of 2-6 month old and their diet was based exclusively on breast milk. Although, according to the American Academy of Pediatrics [43], from the six months onwards, milk is substituted by solid. Other author [44], mentions that breast milk is the primary route where the infants are expose to certain lipophilic toxics that are accumulated by decades in the maternal adipose tissue. If we compare the residues found in this study with the highest levels found in other studies, like the ones reported in [40] for DDE (1.06 mg/kg), and DDT (1.11 mg/kg) in breast milk. The same happened by comparing them with similar works performed in Veracruz with p,p'-DDT and p,p'-DDE (1.27 y 5.02 mg/kg, respectively) [42], and those in the peripheral zone of Mexico City [41], were 108 samples of human milk were analyzed. The content of p,p'-DDT found was 0.117 mg/kg and for p,p'-DDE 2.31 mg/kg. The decrease in the values found was associated to a possible restriction in the use of DDT, although the presence of p,p'-DDE is evident. Specifically, the studies performed in Pueblo Yaqui (Sonora) [45] in breast milk, found levels of p,p'-DDE disturbingly high (6.25 mg/kg), considering that is one of the most important agricultural areas in Mexico. At the present time, in the same zone, the authors reported the presence of p,p'-DDE in 66.66% of the samples of serum from children (between 6 and 12 years of age), with the levels of 0.1 a 443.9 µg/L [46]. These results suggest that DDT is present in the environment and the residues found in biological samples could be due to many factors such as contaminated food consumption. In the present study, the most frequent found metabolite in biological samples was p,p'-DDE due to a its degradation by enzyme system in mammals. According to literature, 50% of p,p'-DDT in the environment could be degrade in 6 years, 67% in 12 years being p,p'-DDE the only product for its degradation [11]. Therefore the contamination present in the studied breast milk can be due to an exposition for more than 12 years. The World Health Organization [47] reports maximum residue limits in foods for DDT and its metabolites of 1.25 mg/kg. While the levels of tolerance established by the FAO/WHO in 1998 [48] for the same compounds in cow milk are 0.05 mg/kg fatty base, in the present study the found levels in breast milk were below the established levels (less than 82%). It is important to mention that DDT levels and other organochlorine compounds in breast milk could be different based on the number of births, age and other factors such as diet, occupation and social status [49, 40]. It is known that levels of these compounds are higher in the breast milk from younger women. In this research, it not was possible to find a correlation

Author details

María-Lourdes Aldana-Madrid[1*], María-Isabel Silveira-Gramont[1],
Fabiola-Gabriela Zuno-Floriano[2] and Guillermo Rodríguez-Olibarría[1]

*Address all correspondence to: laldana@guayacan.uson.mx

1 Departamento de Investigación y Posgrado en Alimentos, Universidad de Sonora. Hermosillo, Sonora, México

2 Department of Environmental Toxicology, University of California, Davis, California, USA

References

[1] Domínguez-Cortinas G, Díaz-Barriga F, Martínez-Salinas RI, Cossio P, Pérez-Maldonado. Exposure to chemical mixtures in Mexican children: high-risk scenarios. Environmental Science Pollution Research. Published online: 29 April 2012. DOI 10.1007/s11356-012-0933-x http://www.ncbi.nlm.nih.gov/pubmed/22544601 (accessed 03 August 2012).

[2] Trejo-Acevedo A, Rivero-Pérez NE, Flores-Ramírez R, Orta-García ST, Varela-Silva JA, Pérez-Maldonado IN. Assessment of the Levels of Persistent Organic Pollutants and 1-hidroxypyrene in Blood and Urine Samples from Mexican Children Living in an Endemic Malaria Area in Mexico. Bulletin of Environmental Contamination and Toxicology 2012;(88) 828-832.

[3] Yañez L, Ortiz-Pérez D, BatreslD, Borja-Aburto VH, Díaz-Barriga F. Levels of Dichlorodiphenyltrichloroethane and Deltamethrin in Humans and Environmental Samples in Malarious Areas of Mexico. Environmental Research 2002;(88) 174-181.

[4] Díaz-Barriga F, Borja-Aburto V, Waliszewski SM, Yáñez L. DDT in Mexico In: Fielder H. (ed.), Persistent Organic Pollutants. The Handbook of Enviromental Chemistry; 2002. p371-388.

[5] Herrera-Portugal C, Franco H, Yáñez G, Díaz-Barriga F. Environmental Pathways of Exposure to DDT for Children Living in a Malarious Area of Chiapas Mexico. Environmental Research 2005;(99) 158-163

[6] Aldana ML, Mendívil CI, Mada CD, Silveira MI, Navarro JL. Alteraciones en el Análisis del Líquido Seminal de Aplicadores de Insecticidas en el Medio Urbano. Acta Médica de Sonora 2003;(4) 5-7

[7] Silveira Gramont MI, Amarillas Cardoza VT, Rodríguez Olibarría G, Aldana Madrid ML, Zuno Floriano FG. Valoración del Riesgo de Exposición a Insecticidas Organofosforados en Adultos del Sexo Masculino en Sonora, México. CIENCIA@UAQ. 2011;4(2) 70-81.

[8] Albert L. Panorama de los plaguicidas en México. Revista de Toxicología en Línea (RETEL) 2005;(8) 1-18. http://www.sertox.com.ar/retel/n08/01.pdf (accessed 02 August 2012).

[9] Rogan W, Chen A, Health Risks and Benefits of bis (4-chlorophenyl)-1,1,1-trichloroethane (DDT). Epidemiology Branch, US National Institute of Environmental Health Sciences 2005;(366) 763-773.

[10] Pérez-Maldonado IN, Trejo A, Ruepert C, Jovel RC, Méndez MP, Ferrari M, Saballos-Sobalvarro E, Alexander C, Yáñez-Estrada L, López D, Henao S, Pinto ER, Díaz- Barriga F. Assessment of DDT Levels in Selected Environmental Media and Biological Samples from Mexico and Central America. Chemosphere 2010;(78) 1244-1249.

[11] ATSDR. Agency for Toxic Substances and Disease Registry. Toxicological Profile for DDT,DDD and DDE. US Department of Health and Human Services. Public Health Service. Atlanta, GA. Registry 2002;1-397. http://www.atsdr.cdc.gov/toxprofiles/tp35-p.pdf (accessed 03 August 2012).

[12] World Health Organization. DDT and its derivatives. Environmental health criteria 9, Geneva: World Health Organization. United Nations Environment Programme. 1979. http://www.inchem.org/documents/ehc/ehc/ehc009.htm (accessed 02 February 2012).

[13] Smith D. Worldwide Trends in DDT Levels in Human Breast Milk. Inf. J. Epidemiol. 1999;(28) 179-188.

[14] Wolff MS, Zoleniuch-Jacquotte A, Dubin N, Toniolo P. Risk of Breast Cancer and Organochlorine Exposure. Cancer Epidemiology, Biomarkers & Prevention 2000;(9) 271-277.

[15] Waliszewski SM, Aguirre AA, Infanzon RM, Silva CS, Siliceo J. Organochlorine Pesticide Levels in Maternal Adipose Tissue, Maternal Blood Serum, and Milk from Inhabitants of Veracruz, Mexico. Archives of Environmental Contamination and Toxicology 2001;(40) 432-438.

[16] Rivero-Rodríguez L, Borja-Aburto VH, Santos-Burgoa C, Waliszewski S. Rios C, Cruz V. Exposure Assessment for Workers Applying DDT to Control Malaria in Veracruz, Mexico. Environmental Health Perspectives 1997;(105) 98-101.

[17] Jaga K, Dharmani C. Global Surveillance of DDT and DDE Levels in Human Tissues. International Journal Occupational Medicine and Environmental Health 2003;(16) 7-20.

[18] Dosemeci M, Alavanja MC, Rowland AS, Mage D, Zahm SH, Rothman N, Lubin JH, Hoppin JA, Sandler DP, Blair A. A quantitative approach for Estimating Exposure to Pesticides in the Agricultural Health Study.Ann Occup Hyp. 2002;46(2) 245-260

[19] Rothlein J, Rohlman D, Lasarev M, Phillips J, Muniz J, McCauley L. Organophosphate Pesticide Exposure and Neurobehavioral Performance in Agricultural and

Nonagricultural Hispanic Workers. Environmental Health Perspectives 2006;114(5) 691–696.

[20] Arcury TA, Quandt SA, Bart DB, Hoppin JA, McCauley L, Grzywacs JG, Robson MG. Farmworker Exposure to Pesticides: Methodology Issues for the Collection of Comparable Data. Environmental Health Perspectives 2006;11(6) 923-928.

[21] Aldana-Madrid ML, Molina-Romo ED, Rodriguez-Olibarria G, Silveira-Gramont MI. Study of organophosphate insecticides and biochemical indicators in blood and urine of urban adult males. American Chemical Society, Division of Agrochemicals. 229th ACS National Meeting, San Diego, CA, USA.PICOGRAM and Abstracts; 2005(68) 71.

[22] Cortés-Genchi P, Villegas-Arrizón A, Aguilar-Madrid G, Paz-Román MP, Maruris-Reducindo M, Juárez-Pérez CA. Síntomas Ocasionados por Plaguicidas en Trabajadores Agrícolas. Revista Médica del Instituto Mexicano de Seguro Social. 2008;46(2) 145-152.

[23] INEGI. Instituto Nacional de Estadística, Geografía e Informática. Anuario Estadístico del Estado de Sonora, México. 2010. http://www.inegi.gob.mx/ (accessed 15 Octuber 2011).

[24] Encinas Kuraica R. La agricultura en los agronegocios en el sur del Estado de Sonora. Thesis. Instituto Tecnológico de Sonora (ITSON). Ciudad Obregón, Sonora, Mexico; 2005.

[25] INEGI. Instituto Nacional de Estadística, Geografía e Informática. Anuario Estadístico del Estado de Sonora, México. 2000. http://www.inegi.gob.mx/ (accessed 08August 2011).

[26] OMS. Organización Mundial de la Salud. Manual de laboratorio de la OMS para el examen de semen humano y de la interacción entre el semen y el moco cervical. Cuarta edición. Ed. Médica Panamericana. Madrid, España; 2001. p11-26.

[27] Dale WE, Miles JW. Quantitative Method for Determination of DDT and DDT Metabolites in Blood Serum. Journal of the AOAC 1970;(53) 1287-1292.

[28] Yagüe C, Bayarri S, Lazaro R, Conchello P, Ariño A, Herrera A. Multiresidue Determination of Organochlorine Pesticides and Polychlorinated Biphenyls in Milk by Gas Chromatography with Electron-Capture Detection after Extraction by Matrix Solid-Phase Dispersion. Journal AOAC International 2001;85(5) 1561-1568.

[29] Pastor Benito S. Biomonitorización citogenética de cuatro poblaciones agrícolas europeas expuestas a plaguicidas mediante el ensayo de micronúcleos. PhD thesis. Universidad Autónoma de Barcelona. Barcelona, España; 2002.

[30] Rojas Companioni D, Rodríguez Díaz T. Monitoreo Biológico para Aplicar a los Trabajadores de una Fábrica de Plaguicidas. Revista Cubana de Higiene y Epidemiología 1998;36(3) 179-184.

[31] Aldana Madrid ML. Efecto bioquímico, histológico y molecular del piretroide ciper-metrina in vivo en presencia y ausencia de α-tocoferol. PhD thesis. Universidad Autónoma de Querétaro, Mexico; 1998.

[32] Rodriguez M, Puga F, Chenker E, Mazanti M. Short-term Effect of Malathion on Rats Blood Glucose and on Glucose Utilization by Mammalian Cells In vitro. Ecotoxicology Environmental Safety 1986;12(2) 110-113.

[33] Ramos LD, Munguía Guerrero L, Tarradellas J. Determinación de Bifenilospoliclorados (BPC's) Residual y Plaguicidas Organoclorados en Peces Comestibles de la Bahía de la Isla de Utila. Contaminantes Químicos/Monografía 4. Tegucigalpa, Honduras; 1994.

[34] Rosser BJ. Australia: uso de plaguicidas en monocultivos de árboles afecta gravemente la salud en Tasmania WRM. Movimiento Mundial por los Bosques Tropicales. Boletín del WRM Número 97; 2005. http://www.wrm.org.uy/boletin/97/Australia.html. (accessed 02 February 2012).

[35] Jiménez Díaz M, Martínez Monge V. Validación de la determinación de colinesterasa plasmática humana a 340 nM. Revista Biomédica 2000;(11) 161-168

[36] Shadnia S, Azizi E, Hosseini R, Khoei S, Fouladdel S, Pajoumand A, Jalali N, Abdollahi M. Evaluation of Oxidative Stress and Genotoxicity in Organophosphorus Insecticide Formulators.Human & Experimental Toxicology 2005;24 (9) 439-445.

[37] Vidyasagar J, Karunakar N, Reddy MS, Rajnarayana K, Surender T, Krishna DR. Oxidative Stress and Antioxidant Status in Acute Organophosphorous Insecticide Poisoning. Indian Journal of Pharmacology. 2004; 36 (2): 76-79.

[38] Pérez V, García-Larios JV, Hernández AF, Gómez MA, Peña GF, Pla A, Villanueva E. Alteración del Perfil Bioquímico Básico en Aplicadores de Plaguicidas en una Zona de Agricultura Intensiva. Utilidad del autoanalizador bioquímico Hitachi 717. Revista Toxicológica 1999;(16) 154.

[39] Villarreal Barrón A. Determinación de plaguicidas organoclorados en suero sanguíneo en niños residentes en el Tobarito, Valle del Yaqui, Sonora. Tesis de Maestría. InstitutoTecnológico de Sonora (ITSON). Sonora, México; 2007.

[40] Prado G, Díaz G, Noa M, Méndez I, Cisneros I, Castoreña F, Pinto M. Levels of Organochlorine Pesticides in Human Breast Milk From Mexico City. Agro Sur 2004;32(2) 60-69.

[41] Prado G, Méndez I, Díaz G, Noa M, González M, Ramírez A, Vega S, Pérez N, Pinto M. Factores de Participación en el Contenido de Plaguicidas Organoclorados Persistentes en Leche Humana en una Población Sub-urbana de la Ciudad de México. Agro Sur 2001;29(2) 128-140.

[42] Waliszewski M, Pardío-Sedas T, Chantiri N, Infanzón M, Rivera J. Organochlorine Pesticide Residues in Human Breast Milk from Tropical Areas in Mexico. Bulletin of Environmental Contamination and Toxicology 1996;(57) 22-28.

[43] AAP. American Academy of Pediatrics. Breastfeeding and Use of Human Milk. Pediatrics 1997;(100) 123-127.

[44] Arcus-Arth A, Krowech G, Zeise L. Breast Milk and Lipid Intake Distributions for Assessing Cumulative Exposure and Risk. Journal of Exposure, Analysis and Environmental Epidemiology 2005;(5) 357-365.

[45] García-Bañuelos L, Meza-Montenegro M. Principales vías de contaminación por plaguicidas en neonatos lactantes residentes en Pueblo Yaqui, Sonora México. ITSON-DIEP 1991;(I) 33-42

[46] Tapia-Quiroz P, Valenzuela-Quintanar A, Grajeda-Cota P, Meza-Montenegro M. Serum levels of organochlorine insecticides in children in the Yaqui Valley, Sonora. Global Environmental Health Workshop, Abstracts. 2007;(19) 40.

[47] FAO/OMS. Food and Agriculture Organization of the United Nations. Residuos de Plaguicidas en los Alimentos. Informe de la Reunión Conjunta 1981 de Expertos de la FAO en Residuos de Plaguicidas y el Medio Ambiente y el grupo de Expertos de la OMS en Residuos de Plaguicidas. Estudio FAO: Producción y Protección Vegetal No. 37. Roma, Italia; 1982.

[48] FAO. Food and Agriculture Organization of the United Nations. Joint meeting of the FAO panel of expert in pesticide residues in food and the environment and the WHO exert group on pesticide residues: Pesticide residues in food: 1963/64-1997 evaluations, Rome; 1998.

[49] Viveros AD, Albert LA, Namihira D. Organochlorine Pesticide Residues in Human Milk Samples from Mexico City. Revista de Toxicología 1989;6(2) 209-221.

[50] Henry RB. Diagnóstico y tratamiento clínico por el laboratorio. Ediciones Científicas y Técnicas, S.A. Novena edición. México; 1997.

Chapter 3

Thiamethoxam: An Inseticide that Improve Seed Rice Germination at Low Temperature

Andréia da Silva Almeida, Francisco Amaral Villela, João Carlos Nunes, Geri Eduardo Meneghello and Adilson Jauer

Additional information is available at the end of the chapter

1. Introduction

The discovery of thiamethoxam has opened new perspectives for the Brazilian agriculture, mainly in seed treatment. The molecule was the center of studies by a group of researchers from official agencies and universities, in order to evaluate its mechanism of action.

Researches were made to establish the activity of the active ingredient on the physiology of the plant, when applied the soybean seed treatment. It was observed that seed germination index and seedling vigor were higher than those of plants in plots without seed treatment. It was also found that, under water stress conditions soybean plants from seed treated with thiamethoxam showed better growth, such as increased length and root volume, faster initial development, higher leaf area, height, number of pods and green colored more intense.

The various pesticides used today can be classified into different classes (CASTRO, 2006), such as:

a. Regulatory Plant or Bio-regulators - organic compounds, non-nutrient, which applied to the plants, at low concentrations promotes, inhibits or modifies some morphological or physiological plant process. The term regulator is restricted to natural or synthetic compounds, applied externally in plants (named exogenous).

b. Plant Hormones - substances produced by plants, which at low concentrations regulate morphological and physiological processes of the plant. Hormones can move within the plant from the generated to the action site, or be produced at the action site. The term hormone is restricted to products that occur naturally in plants (named endogenous).

Belonging to both the previous classes, there are: auxins, gibberellins, cytokinins, abscisic acid and ethylene. It is considered, for the bio-regulator acts, that it must primarily bind to a receptor on the plasmatic membrane of the cell.

c. Plants Stimulants or bio-stimulants - mixtures of plant regulators, occasionally along together nutrients, vitamins, amino acids or miscellaneous debris. They exhibit a different stimulatory effect than if application isolated, creating a synergistic effect between regulators. Some examples of bio-stimulants the Stimulate are Promalin and e mixture GA_3 + 2,4-D.

d. Bioactivators - complex organic substances, that modify the morphology and physiology of plants and are capable of acting in the synthesis and action of endogenous hormones, leading to increase in productivity. In this class some insecticides fit, such as aldicarb and thiamethoxam, besides of the hydrogen cyanamide.

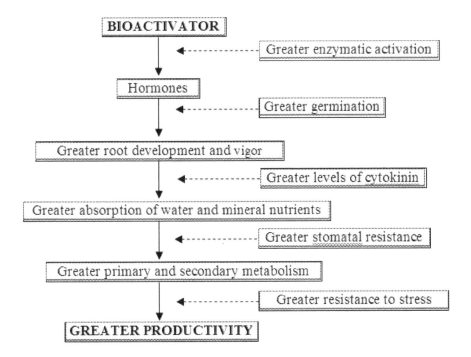

Figure 1. Sequence of events promoted by thiamethoxam (CASTRO, 2006)

Castro (2008), found that the molecule of thiamethoxam is capable of inducing physiological changes in plants. In function of the results obtained, it is concluded that the bioactivator can act in two ways: the first one, is to enable transport proteins from the cell membranes allowing a greater ionic transport, increasing the mineral nutrition of the plant. This increase in the availability of mineral salts promote positive responses in the development and plant productivity (Figure 1). The second one is related to the higher enzymatic activity caused by thiamethoxam, as the seed level or as the plant one. The highest enzymatic activity would increase both the primary and the secondary metabolism. It would increase the synthesis of amino acids, precursors of new proteins. The plant response to these proteins and hormone biosynthesis could be related to important increases in production (Figure 2).

The bioactivators are organic substances, potentially modifying the morphology and physiology of plants, by acting on the synthesis and action of endogenous hormones and may lead to increases in productivity.

In general, insecticides and fungicides are used to control insects and fungi, respectively. However, it has been found that certain chemicals may also exert actions modifying the morphology and metabolism of plants.

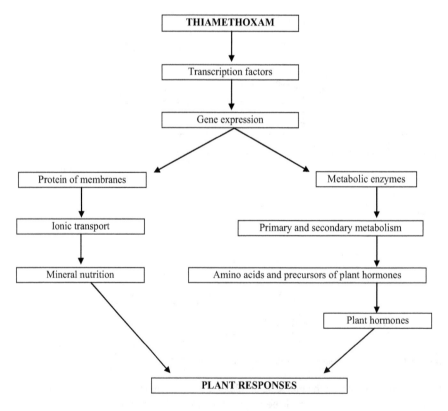

Figure 2. Action mode of thiamethoxam in plants (CASTRO, 2006)

Certain insecticides like aldicarb, carbofuran and thiamethoxam, may cause a physiological effect promoting changes in certain processes in plant physiology, such as growth, morphology or plant biochemistry.

The thiamethoxam can be applied in seed treatment, by spraying on leaves of plants or by soil application, being absorbed by the roots. Applied as a seed treatment, the thiamethoxam can promote the expression of the effect by stimulating root growth and increasing germination rate, consequently, reducing the time for field crop establishment.

2. Physiological changes in rice seeds exposed to low temperature at germination

Rice is grown in diverse environmental conditions, but when compared to other cereals such as oats or wheat, is much more sensitive to low temperatures (Mertz et al., 2009). The occurrence of cold weather is one of the major problems when irrigated rice in Rio Grande do Sul - Brazil, is cultivated since the most of the cultivars in use are from tropical origin. The occurrence of low temperatures, together with the susceptibility of the materials used can cause serious damage to the establishment of the crop, reducing the initial stand and consequently favoring the establishment of weeds. The productivity of irrigated rice in Rio Grande do Sul has suffered strong oscillations over the years, caused in part by climatic conditions, where the occurrence of low temperatures has been one of the major determinants factors of this variability at the productivity levels (Mertz et al., 2009).

On the other hand, hormone controllers have received increasingly more attention in agriculture as the crop techniques develop, especially in high value crops. The bioactivators are complex organic substances that can alter the growth, capable to act on the transcription of DNA in plant, gene expression, membrane proteins, metabolic enzymes and mineral nutrition (Castro and Pereira, 2008). The thiamethoxam insecticide has shown positive effects such vigor expression increase, biomass accumulation, high photosynthetic rate and deeper roots (Cataneo, 2008).

The aim of this work was to evaluate the influence of thiamethoxam in the rice crop and the potential benefits that treatment can provide, when rice seeds are subjected to low temperature during germination and emergence.

3. Material and methods

Three rice cultivars where used: two conventional (BR IRGA 417, BR IRGA 424) and one hybrid (Avax R.). The cultivars had the same physiological quality and were evaluated for tolerance to low temperature through the germination test. The seeds were treated with a commercial product containing 35 grams of thiamethoxam active ingredient per liter of product. The treatments were: Treatment 1 - untreated seeds; Treatment 2 - 100ml of product/100kg of seed; Treatment 3 - 200 ml of product/100kg of seed; Treatment 4 - 300 ml of product/100kg of seed and Treatment 5 - 400 ml of product/100kg of seed, prior to sowing.

The germination test was performed in three replications, eight sub-samples of 50 seeds (400 seeds per replicate) for each cultivar. The seeds were placed to germinate in paper rolls moistened with water equivalent to 2,5 times the weight of the substrate, following the criteria established by the Rules for Seed Testing (Brazil, 2009). Five germination temperatures were used: 25, 20, 18, 15 and 13°C. The germination test at temperatures of 25 and 20°C were performed in the germinator, and at temperatures 18, 15 and 13°C held in BOD. The counting of normal seedlings was performed seven days after sowing for temperatures of 25, 20 and 18°C and at 21 days for temperatures of 15 and 13°C.

4. Results and discussion

According to the results, the rice seeds cultivars BR IRGA 417, BR IRGA 424 and Avax R. treated with thiamethoxam, were superior in all tested temperatures, when compared to the values obtained in the zero dose (without application of thiamethoxam), varying only the intensity of this difference due to the dose used and the temperature.

By observing the data shown in Figure 3, it is found that the treated seeds showed significant increases in germination at different temperatures.

The temperatures of 15°C and 13°C were the most adverse ones, but when the seeds are treated independent from the dose, they showed germination over the zero dose. At the dose of 200 mL/100 kg of seeds at a temperature of 15°C, there was an increase of 21 percentage points, whereas at 13°C this increase was 37 percentage points. At temperatures of 25, 20 and 18°C this increase was on average 7 percentage points when compared with the zero dose.

Figure 4 shows that seeds treated with thiamethoxam at different temperatures had positive additions in relation to the zero dose. The results of this study confirm those obtained by Castro et al. (2007), working with soybeans, and those by Clavijo (2008) working with rice, when claiming that seeds treated with thiamethoxam had their germination accelerated by stimulating the enzymes activity, besides of showing more uniform emergency and stand more uniform and better initial impulse. Also in soybean seeds, Cataneo (2008) observed that thiamethoxam accelerates germination, and induces further development of the embryonic axis. According to the results, rice seeds, cultivars BR IRGA 417, BR IRGA 424 and Avax R. treated with thiamethoxam, were superior in all the tested temperatures, when compared to the values obtained at the zero dose (without application of thiamethoxam), varying only the intensity of this difference due to the dose used and temperature.

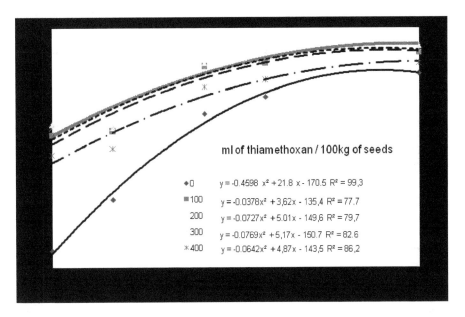

Figure 3. Germination (%) rice seeds, cultivar BR IRGA 417, treated with thiamethoxam at different temperatures.

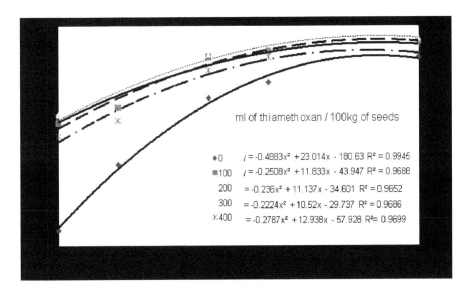

Figure 4. Germination (%) rice seeds, cultivar BR IRGA 424, treated with thiamethoxam at different temperatures.

According to Figure 5, the results of cultivar AVAXI R, hybrid rice seeds when treated with thiamethoxam showed increases in relation to the dose zero. The dose 100mL/100 kg of seeds showed higher increases when compared with other doses at all studied temperatures, being of 28 percentage points at a temperature of 13°C which is the most drastic one, comparing the doses 100mL/100kg of seeds with the zero dose.

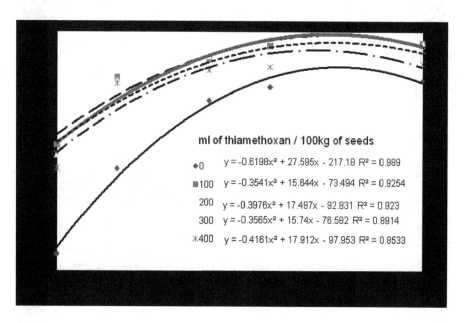

ml of thiamethoxan / 100kg of seeds

◆0 $y = -0.6198x^2 + 27.595x - 217.18$ $R^2 = 0.989$

■100 $y = -0.3541x^2 + 15.644x - 73.494$ $R^2 = 0.9254$

200     $y = -0.3976x^2 + 17.497x - 92.831$ $R^2 = 0.923$

300     $y = -0.3565x^2 + 15.74x - 76.582$ $R^2 = 0.8914$

✳400 $y = -0.4161x^2 + 17.912x - 97.953$ $R^2 = 0.8533$

Figure 5. Germination (%) rice seeds, cultivar Avax R., treated with thiamethoxam at different temperatures.

It was observed that at all temperatures studied with product addition there was an increase in germination of the seeds (Figure 6). In average there were increases reaching up to 8 percent germination at 25 º C, 12 percentage points at 20 º C, 17 percentage points at 18 º C and 34 percentage points in the germination test with temperatures of 15 and 13 ° C when compared with seedlings from untreated seeds. Besides increasing the percentage of germination, there is also the activating effect of the product, with the increase in size of roots and shoots (Figure 7). This increase may provide a more rapid and uniform establishment of the crop. According to Clavijo (2008), the thiamethoxam is transported inside the plant through its cells and activates several physiological reactions like protein expression. These proteins interact with various mechanisms of defense related to the plant stresses, allowing to a better deal with adverse conditions such as drought, low pH, high soil salinity, free radicals, stress by high or low temperature, toxic effects of high levels of aluminum injury caused by pests, winds, hail, attack of viruses and nutrient deficiency

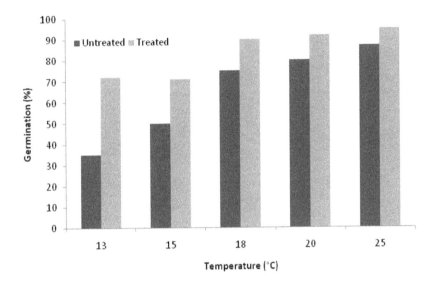

Figure 6. Average germination, cultivar BR IRGA 417, under different doses of thiamethoxam.

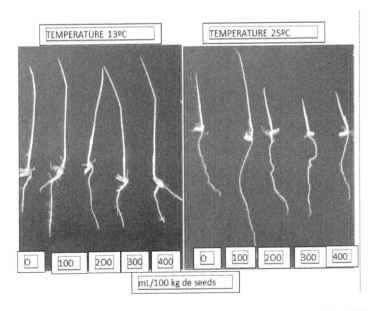

Figure 7. Growth of rice seedlings exposed to different doses of thiamethoxam temperatures of 13 and 25°C

5. Conclusion

The rice seed treatment with thiamethoxam positively favors the physiological quality of seeds.

The doses of 100 and 200 mL of product per 100 kg of rice seed are more effective to improve the physiological performance of rice seeds, in temperatures between 13 and 25 °C.

Author details

Andréia da Silva Almeida[1], Francisco Amaral Villela[2], João Carlos Nunes[3], Geri Eduardo Meneghello[4] and Adilson Jauer[5]

1 Seed Care Institute- Syngenta, Brazil

2 PPG Ciência e Tecnologia de Sementes, Universidade Federal de Pelotas, Brazil

3 Syngenta Crop Protection- Seed Care Institute, Brazil

4 Universidade Federal de Pelotas, Brazil

5 Syngenta Crop Protection, Brazil

References

[1] Brasil. Ministério da Agricultura e Reforma Agrária. Regras para análise de sementes.   http://www.agricultura.gov.br/images/MAPA/arquivos_portal/ACS/sementes_web.pdf

[2] Cataneo, A C. Ação do Tiametoxam (Thiametoxam) sobre a germinação de sementes de soja (*Glycine max*.L): Enzimas envolvidas na mobilização de reservas e na proteção contra situação de estresse (deficiência hídrica, salinidade e presença de alumínio). Tiametoxam: uma revolução na agricultura brasileira.: Gazzoni, D.L. (Ed.)., 2008, p. 123-192.

[3] Clavijo, J. Tiametoxam: um nuevo concepto em vigor y productividad. Bogotá, Colômbia, 2008.196p.

[4] Castro, P.R.C. Agroquímicos de controle hormonal na agricultura tropical. Piracicaba: ESALQ, 2006. 46p. (Série Produtor Rural, 32).

[5] Castro, P. R. C.; Pereira, M.A. . Bioativadores na agricultura. Tiametoxam: uma revolução na agricultura brasileira.: Gazzoni, D.L. (Ed.). 2008, p. 118-126.

[6] Castro, P. R. C.; Pitelli, A.M.C.M.; Peres, L.E.P.; Aramaki, P.H. . Análise da atividade reguladora de crescimento vegetal de tiametoxam através de biotestes. Publicatio. UEPG (Ponta Grossa), v. 13, p. 25-29, 2007.

[7] Mertz, L.M.; Henning, F.A.; Soares, R.C.; Baldiga, R.F.;Peske,F.B.; Moraes, D.M.Alterações fisiológicas em sementes de arroz expostas ao frio na fase de germinação. Revista Brasileira de Sementes, v. 31, n. 2. p 262-2701. 2009.

Chapter 4

Spatial and Monthly Behaviour of Selective Organochlorine Pesticides in Subtropical Estuarine Ecosystems

T.S. Imo, T. Oomori, M.A. Sheikh, T. Miyagi and
F. Tamaki

Additional information is available at the end of the chapter

1. Introduction

Organochlorine pesticides (OCPs) are one of the most important persistent organic pollutants (POPs) which pose threats to ecosystems and human health. The twelve so-called POPs; nine of which are organochlorine pesticides. The two organochlorine pesticides (HCH and dieldrin) were used in or arise from industry mainly for agriculture purposes. Sediments serve as both a source and a removal mechanisms for contaminants to and from rivers and streams and as a means of contaminant transport downstream. Sediment also provides habitat for benthic biota and can be in the food web around rivers and stream, and some organisms such as fish are consumed by people and birds (Brasher & Anthony, 1998; Laabs et al., 2002). Although the residue levels of the chlorinated compounds in the environments have considerable declined in the past 20 years, recent work has depicted that chlorinated pesticides could be detected in the range of 0.03-25.17 ngg^{-1} (dry weight) (Chang & Doong, 2006; Zhou et al., 1994).

Some OCPs such as DDT and endosulfan are still used in some countries around the tropical and subtropical regions for agricultural and medicinal purposes. These compounds can be deposited into the sediments through long-range atmospheric transport, resulting in a high exposure to OCPs in the area near the pollution source (Tanabe et al., 1994; Doong et al., 2002; Fabricius, 2005). River bed sediments and fish tissues contain higher concentrations of organochlorine compounds than the surrounding water, so analysis of sediment increases the likelihood of detecting compounds that are present in the river.

1.1. A brief history of organochlorine pesticides (OCPs)

A beginning of the twentieth century, early research and studies with chemical pesticides led to the widespread use of inorganic compounds within agriculture containing elements such as sulphur, arsenic, mercury, lead and other metals (Turnbull, 1998). For some natural products such as pyrethrum were also known to be effective pesticides at the time, but were considered to expensive for widespread use (Awofolu and Fatoki, 2003). Between the world wars, the development of the chlor-alkali industry provided the raw material for the mass production of synthetic chlorinated organic molecules. The first and early chlorinated phenoxy acid herbicide (2,4-D) was first discovered in 1932 (Burton and Bennett, 1987). Although this chemical rapidly breaks down in the environment, the seed fungicide hexachlorobenzene (HCB) were introduced in 1933 was found to be far more persistent (Carlsen et al., 1995). The structurally similar insecticides hexachlorocyclohexane or HCH or BHC (also known as benzene hexachloride-BHC) also emerged at this time. The outbreak of war in 1939 and the need to administer malaria and typhus amongst soldiers and civilians has led to the uncovering, unravelling and application of DDT across the world within four and half years from 19430. Related research about the nerve gas agents in Germany led to the discovery of the associated organophosphorus pesticides (Carlsen et al., 1995). Towards the end of the world war, a clear new future for the agrochemical control using these organochlorine chemicals was contemplated. After the world was, the British government considered a practical need to improve agricultural activity and increase food production by the admittance of more complex machinery creation of larger fields, use of chemical fertilizers and the new synthetic pesticides. By 1953, two insecticidal seed dressings, dieldrin and aldrin were being introduced into the UK (Burton and Bennett, 1987). In America, toxaphene was first produced in 1945 as an effective insecticide for cotton plants. This mixture of over 170 chlorinated derivatives known as camphachlor in Europe was recommended as an alternative to DDT before it was banned in the 1980s due to its environmental toxicity (Carlsen et al., 1995). Coupled with other persistent organochlorines such as chlorofluorocarbons, chlorinated biphenyls, dibenzodioxins and dibenzofurans (Doong et al., 2002) the chlorinated pesticides have the potential to cause significant damage to the natural ecosystem by interfering with reproductive processes, this influencing the biodiversity of non-target organisms (Forget et al., 1995). Some aspects of this impairment are now well researched and documented. Whether from past application in developing countries or from continuing current use, these compounds can now be detected in the most remote regions of the planet.

1.2. Definition and importance of organochlorinepesticides (OCPs)

Organochlorines are carbon-based chemicals that contain bound chlorine. These compounds are hydrophobic and lipophilic to varying degrees, meaning their solubility in water is very low, whereas their solubility in fats and oils is relatively high (Cheevaporn et al., 2005). They are noted for their persistence and bioaccumulation characteristics. Some substances may be very persistent in the environment (i.e. with half-lives ($t_{\frac{1}{2}}$) greater than 6 months). The nature of this persistence needs to be clarified- it is the length of time the compound will remain in the environment before being broken down or degraded into other and less hazardous substances (Forget et al., 1995). The widespread use of these compounds over the past half century has led to their detection in many hydrologic systems world-wide from agricultural and non-agricultural purposes (Monirith et al., 2003). Organochlorine pesticides (OCPs) are considered to be dangerous not only for the environment but for animals and human beings as well. They are very stable substances and it has been cited that the degradation of DDT in soil in 75-100% in 4-30 years (Doong et al., 2002). Other chlorinated pesticides such as Aldrin, Dieldrin, Endrin and Isodrin remain stable in water for many years after their use (Cheevaporn et al., 2005).

1.3. Chemistry

Organochlorine pesticides (OCPs) are organic compounds that highly resistant to degradation by biological, photolytic or chemical means. OCPs are mostly chlorinated. The carbon-chlorine bond is very stable towards hydrolysis and the greater number of chlorine substituted and functional groups, the greater the resistance to biological and photolytic degradation (Doong et al., 2002). Chlorine attached to an aromatic (benzene) ring is more stable to hydrolysis than chlorine in aliphatic structures (Forget et al., 1995). As a result, OCPs are typically ring structures with a chain or branched chain framework. By virtue of their solubility leading their propensity to pass readily through the phospholipids structure of biological membranes and accumulate in fat deposits.

1.4. Human health

As noted for environmental effects, it is also most difficult to establish cause and effects relationships for human exposure of OCPs and incident diseases. As with wildlife species, human encounter a broad range of environmental exposures and frequently to a mixture of chemicals at any time. Much work remains to be done on the study of the human health impact of exposure to OCPs, particularly in view of the broad range of concomitant exposing experienced by humans (Vagi et al., 2005). Previous and present scientific evidences suggest that some OCPs have the potential to cause significant adverse effects to human health at the local level and at the regional and global levels through long-range transport (Doong et al., 2002). For some OCPs, occupational and accidental high-level exposure is of concern for both acute and chronic worker exposure. The risk is greatest in developing countries where the OCPs in tropical agriculture have resulted in a large number of deaths and injuries (Fu et al., 2003). In addition, to other exposure courses, workers exposure to OCPs during waste management is a significant source of high concentration of certain OCP which resulted in illness and death (Doong et al., 2002). For example, a study in the Philippines showed that in 1990, endosulfan became the number one cause of pesticide-related acute poisoning among subsistence rive-farmers and mango sprayers (Forget et al., 1995). Earliest reports of exposures to OCPs related to human health impact include an episode of HCB poisoning of food in south-east Turkey, resulting in the death of 90% of those affected and in other exposure related incidences of hepatic cirrhois, porphyria and urinary arthritic and neurological disorders (Barakat, 2004). Occupational, bystanders and near field exposure to toxic chemicals is often difficult to minimize in developing countries (WHO, 2004). Laboratory and field observations on animals as well as clinical demonstrate that over exposure to certain OCPs may be associated with a wide range of biological effects. These adverse effects may include immune dysfunction, neurological deficits, reproductive anomalies, behavioural adnormalities and carcinogenesis (Forget et al., 1995). The scientific evidence demonstrating a link between chronic exposure to sub lethal concentrations of OCPs (such as that which could occur as a result of long range-transport) and human health impacts is more difficult to establish but gives cause for serious concern (Doong et al., 2002).

1.5. Organochlorine pesticides production and use

The nine OCPs out of twelve POPs compounds were used in or arise from industry, agriculture crops and disease vector control of public health (Chang and Doong, 2006). By the late 1970 all eight OCPs has been either banned or subjected to severe use restriction in the developed world but the major release of these compounds were mostly used by developing countries especially Asia (Hung and Thiemann, 2003), South/Central America (Falco et al., 2003) and Africa (Mwevura et al., 2002). Although the statistics on the use in many areas remained unclear (FAO, 1989). Previous studies revealed that some HCH remains a common compound used in large quantities in India, China, Africa and South America (Turnbull, 1995). It was also recorded that India consumed 25,000 tons of HCH annually over recent years (Davis et al., 1992) and one factory in China was thought to have an annual product of 20,000 tons (Zhang et al., 2002). In Japan, the using of these OCPs has been prohibited in the field in the 1970-1980 (Nakai et al., 2004). It was estimated that the pesticide used in the United States was 550,000 tons during 1995 (Golfinopoulos et al., 2003). In addition, Greece consumed approximately 3500 tons per year of OCPs in the form of insecticides and pesticide (Miliadis, 1993). In Vietnam, approximately 15,000 tons was used from 1957-1972 (Quyen et al., 1995) and 50 tons from the year 1999 (Hung and Thiemann, 2003). In Germany, 36,000 tons was consumed in the year 1991 (Statistisches Bundesamt, 1993, Hung and Thiemann, 2003).

1.6. Characteristics of hexachlorocyclohexane (HCH)

Hexachlorocyclohexane (HCH) is an insecticide that exists in eight different forms. One of its form is known as gamma-HCH (γ-HCH) or commonly called Lindane is produced and used as an insecticide on fruit, vegetables and forest crops. It is a white solid that turns into a vapour when released into the air with a melting point varied with isomeric composition. Its vapour pressure at 4.2 mm Hg at 20°C. (US EPA, 2012).

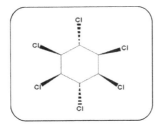

Figure 1. Chemical structure of hexachlorocyclohexane

1.7. Characteristics of dieldrin

Dieldrin is an insecticide which is closely related to aldrin, which reacts further to form dieldrin. It used principally to control textile pests and insects living in agricultural soils. It is a white crystals with a melting point of 175-176°C. Its solubility in water is 140 µgL^{-1} at 25°C with a vapour pressure of 1.78×10^{-7} mm Hg at 20°C (US EPA, 2012).

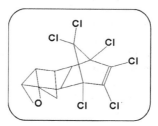

Figure 2. Chemical structure of dieldrin

2. Environmental impact to estuaries

Organochlorine pesticides are carbon-based chemicals that contain bound chlorine. These compounds are hydrophobic and lipophilic to varying degrees, meaning their solubility in water is very low, whereas their solubility in fats and oils is relatively high (Cheevaporn et al., 2005).They are noted for their persistence and bioaccumulation characteristics. The widespread use of these compounds over the past half century has led to their detection in many hydrologic systems world-wide from agricultural and non-agricultural purposes (Monirith et al., 2003). The presence of HCH and dieldrin pesticides in the environment may be related to both past and present land use in a watershed. It enters the aquatic environment from a variety of sources, including the atmosphere, industrial and municipal effluents and agricultural and urban non-point source run-off. HCH and dieldrin are mostly associated with bottom sediments, which can be ingested by benthic organisms. These organisms are then eaten by fish and birds, which can result in higher concentrations through aquatic and terrestrial food chains. Due to the long residence time of these substances in the environment, it is important to examine the pollution they cause not only the environment but also for the lower invertebrates such as corals. Since the ocean is the receiving basin for terrigenous freshwater run-off and its entrained materials, some fractions of these compounds that are used in upland eventually reach the marine ecosystems.

The Manko and Okukubi estuaries are protected wetlands located in a subtropical climate on Okinawa Island. These estuaries are very famous host for migrating birds from South East Asia and mainland Japan. It also plays a great role of species conservation and it was added to the RAMSAR Convention register of wetlands. However, estrogenic activities were detected in sediment samples from these estuaries (Tashiro et al., 2007). Previous studies showed that the coral reef ecosystems and their adjacent environments in and around the Okinawa Island are contaminated with OCPs, OTCs and PCBs (Tashiro et al., 2003; Imo et al., 2007; Sheikh et al., 2002). However, very little is known about the behaviour of HCH and dieldrin in estuarine sediments of subtropical areas. The main objective of this chapter is provide crucial information on the distribution and behaviour of HCH and dieldrin compounds in protected subtropical estuaries in the Okinawa Island.

3. Experimental

3.1. Sample processing

Surface sediment samples were collected with a stainless steel grab. The upper 1-3 cm of the sample were carefully removed and stored in acid rinsed polyethylene 250 mL glass bottles. Samples were transferred to the laboratory and were stored at -20ºC until sample extractions. Details of sampling areas is shown in Table 1 and description of samples as shown in Table 2. Sampling location is shown in Figure 3.

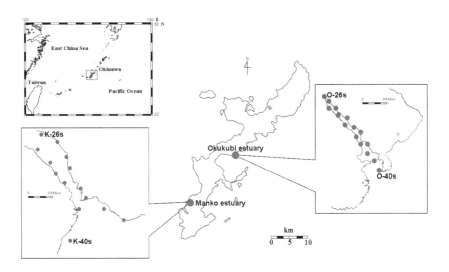

Figure 3. Sampling locations

Estuary	Transect	Sample	Location	Activities
Manko	TM1	K-26s-05		
		K-27s-05	Upstream	Residential area
		K-28s-05		
	TM2	K-29s-05		
		K-30s-05	Mid –stream	Residential area
		K-31s-05		
	TM3	K-32s-05		
		K-33s-05	Mid-stream	Residential area
		K-34s-05		
	TM4	K-35s-05		
		K-36s-05	River mouth	Residential and fishing port
		K-37s-05		
	TM5	K-38s-05		
		K-39s-05	Naha port	Commercial port
		K-40s-05		
Okukubi	TO1	O-26s-05		
		O-27s-05	Upstream	Agriculture
		O-28s-05		
	TO2	O-29s-05		
		O-30s-05	Mid-stream	Agriculture
		O-31s-05		
	TO3	O-32s-05		
		O-33s-05	Mid-stream	Agriculture
		O-34s-05		
	TO4	O-35s-05		
		O-36s-05	Mid-stream	Fishing area
		O-37s-05		
	TO5	O-38s-05		
		O-39s-05	River mouth	Residential, public area
		O-40s-05		

Table 1. Details of sampling areas.

Sample	Surface river sediments
	Sample sketch
K-26s-05	mud
K-27s-05	mud
K-28s-05	mud
K-29s-05	mud
K-30s-05	mud
K-31s-05	mud
K-32s-05	mud
K-33s-05	mud
K-34s-05	mud-sandy
K-35s-05	sandy
K-36s-05	sandy
K-37s-05	sandy
K-38s-05	sandy
K-39s-05	sandy
K-40s-05	sandy
O-26s-05	sandy
O-27s-05	sandy
O-28s-05	sandy
O-29s-05	sandy
O-30s-05	sandy
O-31s-05	sandy
O-32s-05	sandy
O-33s-05	sandy
O-34s-05	sandy
O-35s-05	sandy
O-36s-05	sandy
O-37s-05	sandy
O-38s-05	sandy
O-39s-05	sandy
O-40s-05	sandy

Table 2. Description of surface sediment samples

3.2. Sample extraction

Prior to extraction, surface sediments were freeze-dried, homogenized with a stainless spatula and passed through a 63 μm sieve followed by mixing with anhydrous Na_2SO_4. The sediment samples were extracted by ultrasonication technique as described by (Vagi et al., 2005). The surrogate standard of 5 μgL^{-1} was added to 40 g of sediments. Portion of this amount was used for QC analysis (i.e. each batch contained 1 sample, 1 blank, 4 spiked). For the spiked samples, various concentration of Chlorinated Mix (5, 10, 50 μgL^{-1}) were added to each spiked sample. The samples were extracted by ultra wave sonication for 15 minutes with 10 mL of dichloro-methane. The extracts were then filtered using WHATMAN filters (0.45 μm) followed by centrifugation at 3000 rpm for 15 minutes. The clear organic supernants were removed then the combine extracts were evaporated on a rotary evaporator at 30-35°C near to dryness. A 1 mL of hexane was added to the dried residues. For further cleanup, the samples were then added to the florisil ENVI Carb and the samples were eluted with hexane. The residues were dissolved in 1 mL hexane. A 1mL of Internal Standard (Pentachloronitrobenzene, 50 μgL^{-1}) in the amount extracted before GC-MS analysis.

3.3. Environmental parameters

3.3.1. Total Organic Carbon (TOC) in sediments

Approximately 3g sediment sample was weighed (± 0.002 g) and HCl (*2M*) was added to the sample and left over night to remove all carbonates. Milli Q water was added to rinse the acid from the sediments. To ensure that all the acid was removed from the samples, a *6M* of HCl was added. The acid from the sample was removed by adding 2 mL of distilled water followed by centrifugation. The sediments were then dried at 60°C over night and ready for analysis (US EPA, 2012). The Total Organic Carbon (TOC) was determined using the CHNS analyser (JM 10 Model from J-Science Lab, Co. Ltd, Japan0. Calibration was performed using Antipyrine as Standard with the following compositions: C_8H_9NO = 135.17 (C = 70.19%, H = 6.43%, O = 8.50% and N = 14.88 %).

3.3.2. pH

The pH was measured using a portable pH meter at room temperature in the laboratory using PHM 95/ion meter, Radiometer model (± 0.001 pH).

4. Results and discussion

The highest concentration of HCH was found in the sampling month of October (213 ng/g) (Manko estuary) and Dieldrin (98 ng/g) (Okukubi estuary) followed by the month of November (HCH-199 ng/g) (Manko estuary) and (Dieldrin-90 ng/g) (Okukubi estuary). The status of HCH and dieldrin in sediments in this study was compared with those in other rivers. The levels of OCPs in this study are lower than that of Er-jen river, Taiwan (80-8200 ng/g (dw))

(Zhang et al., 2002). River Mataniko, Solomon Island (140 ng/g (dw)) (Iwata et al., 1995) but higher in some rivers in Japan (2.5-12 ng/g (dw)) (Sakar et al., 1997). The basic physico-chemical parameters of sediments such as TOC were also measured. The TOC contents ranged from nd-3.96% (Table 3 and Table 4). Figure 4-Figure 5 shows a positive correlation with the concentration of HCH especially in the Manko estuary. No correlation was shown between HCH and TOC from the Okukubi estuary. It is clear that sediments from the Okukubi estuary were composed of fine particles. This observation is consistent with other studies which demonstrated that fine particles can retain large amounts of organic compound and pose a high pollution potency (Hong et al., 1995). Since HCH and dieldrin exhibit carcinogenic activities, the contamination levels detected may pose a high ecotoxicity for aquatic and marine organisms.

Sample	Temperature (°C)	pH	Total Organic Carbon (%)
K-26s	26.5	8.01	3.13
K-27s	27.4	7.86	2.74
K-28s	28.4	7.81	1.29
K-29s	27.9	7.43	0.24
K-30s	27.6	7.56	nd
K-31s	27.6	7.66	nd
K-32s	28.9	7.74	nd
K-33s	28.9	7.65	nd
K-34s	28.2	7.55	nd
K-35s	28.2	7.64	nd
K-36s	28.6	7.62	nd
K-37s	28.6	7.58	nd
K-38s	28.9	7.56	nd
K-39s	28.9	7.58	nd
K-40s	28.9	7.56	nd

nd: not detected

Table 3. Summary of Environmental Parameters – Manko Estuary

Sample	Temperature (°C)	pH	Total Organic Carbon (%)
O-26s	25.4	8.06	3.96
O-27s	25.9	8.00	2.82
O-28s	26.0	7.99	2.11
O-29s	26.2	7.99	1.76
O-30s	26.5	7.78	0.54
O-31s	26.9	7.56	nd
O-32s	27.0	7.69	nd
O-33s	27.2	7.69	nd
O-34s	27.0	7.72	nd
O-35s	27.1	7.71	nd
O-36s	27.1	7.60	nd
O-37s	26.8	7.64	nd
O-38s	26.9	7.64	nd
O-39s	26.9	7.58	nd
O-40s	26.9	7.54	nd

nd: not detected

Table 4. Summary of Environmental Parameters – Okukubi Estuary

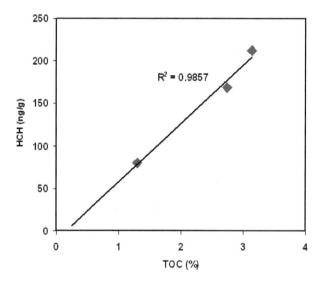

Figure 4. Correlation of HCH with TOC [Manko estuary]

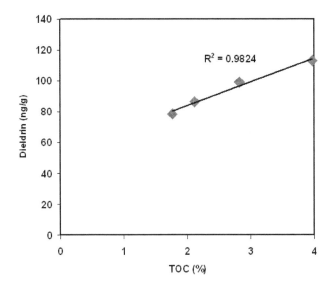

Figure 5. Correlation of Dieldrin with TOC [Okukubi estuary]

4.1. Spatial distribution of HCH and dieldrin in sediments

The highest concentrations of HCH was found in sample K-26s – 213 ng/g (dw) (Manko estuary) and the highest dieldrin concentration was found in sample O-26s – 98 ng/g (dw) (Okukubi estuary). Most samples in the Okukubi estuary had relatively low levels of HCH compared to Manko estuary, where the sediments mainly composed of sand. It may be due to the similar historical input and deposit indicating important sources of these organochlorine pesticides in these areas. The second highest concentration of HCH in the Manko estuary, 99 ng/g (dw) followed by 90 ng/g (dw). The second highest concentration of dieldrin in the Okukubi estuary was 199 ng/g (dw) followed by 95.5 ng/g (dw). The levels of HCH and dieldrin in this study are higher in those found in the sediments of the Mingjiang River Estuary, China (2.99–16.21 ng/g, with a mean value of 8.62 ng/g dw (Kennicutt et al., 1994). the Wushi Estuary, Taiwan (0.99–14.5 ng/g, with a mean value of 3.78 ng/g dw (Iwata et al., 1995) Xiamen Harbor,

China (0.14–1.12 ng/g, with a mean value of 0.45 ng/g dw, Hong et al., 1995) and Casco Bay, USA (<0.25– 0.48 ng/g) dw, but lower than the Matanico River and Solomon Islands (140 ng/g) (Walker et al., 1999).

4.2. Monthly variations of HCH and dieldrin in sediments

It clearly revealed that the HCH and dieldrin pesticides residues in October were higher compared to other sampling months. This means that some organochlorine pesticides could be released from the run-off effluents to waters with much rainfall during the rainy season and typhoon season in Okinawa during the summer. In all sampling months, the highest concentration of organochlorine pesticides in the Manko estuary was 213 ng/g (dw) in October and for the Okukubi estuary, the highest concentration of organochlorine pesticides was also detected in the month of October, 213 ng/g (dw). The second highest concentration of HCH was detected in the month of November (199 ng/g (dw)) followed by the month of December (99 ng/g (dw)) for the Manko estuaries. The second highest concentration of dieldrin was detected in the month of October (90 ng/g (dw)), followed by 89.5 ng/g (dw) in the month of November for the Okukubi estuaries.

4.3. Composition analyses in sediments

Composition difference of HCH in the environment could indicate contamination sources (Wu et al., 1999). Technical HCH has been used as broad spectrum pesticides for agricultural purposes, which has been banned in the 1970's in Japan. Technical-grade HCH consists principally of four isomers, α-HCH, β-HCH, γ-HCH and δ-HCH. The physiochemical properties of these HCH isomers are different. The β-HCH has the lowest water solubility and vapour pressure which is the most stable and relatively resistant to microbial degradation (Strandberg et al., 1998). Also it should be noted that α-HCH can be converted to β-HCH in the environment (Lee et al., 2001). The results showed that a high percentage of HCH isomer was recorded in the sampling months December, January and February. It is possible that HCH may be re-absorbed to surface sediments. There was no strong evidence to prove the recent usage of HCH in Okinawa; however Manko estuary was contaminated with HCH.

5. Conclusion

Generally the distribution of organochlorine pesticides were associated with land use practices including agriculture and urbanization and the sediments from estuary have higher contents of organic matter such as TOC and organochlorine pesticides residues. The concentration and compositions of organochlorine pesticides varied significantly with different sampling sites. The HCH in the surface sediments were well correlated with TOC content. The organochlorine pesticides residues (HCH and dieldrin) were detected due to re-absorption in sediments due to previous deposition. The possible sources of these organochlorine pesticides are still unknown but they may come from residential areas, commercial and naval ports and agriculture activities.

Author details

T.S. Imo[1,5], T. Oomori[1], M.A. Sheikh[1,4], T. Miyagi[2,3] and F. Tamaki[3]

1 Department of Chemistry, Graduate School of Engineering and Science, University of the Ryukyus, Nishihara, Okinawa, Japan

2 Okinawa Prefectural Institute of Health and Environment, Ozato Ozato Nanjo, Okinawa, Japan

3 Water Quality Control Office, Okinawa Prefectural Bureau, Miyagi, Chatan, Okinawa, Japan

4 Research Unit, The State University of Zanzibar, Tanzania

5 Faculty of Science, National University of Samoa, Samoa

References

[1] Awofolu R.P and Fatoki O.S (2003). Levels of Cd, Hg and Zn in some surface waters from the Eastern Cape Province, South Africa. Water SA 29,4, 375-380

[2] Barakat A.O (2004). Assessment of persistent toxic substances in the environment of Egypt. Environmental International 30, 309-322

[3] Brasher A.M and Anthony S.S. Occurrence of Organochlorine pesticides in stream bed sediment and fish from selected streams on the Island of Oahu, Hawaii. 1998, USGC. Factsheet 140-00

[4] Burton M.A.S and Bennett B.G (1987). Exposure of man to environmental hexachloro-benzene (HCB) – an exposure commitment assessment. Science and Total Environment 66, 137

[5] Carlsen E, Giwercman A, Keiding N and Skakkerbaek N.E (1995). Assaying estrogenicity by quantitating the expression levels of endogenous estrogen-regulated genes. Environmental Health Perspectives 103, 7: 137

[6] Chang S and Doong R. Concentration and fate of persistent organochlorine pesticide in estuarine sediments using headspace solid-phase microextraction 2006. Chemosphere 62, 1869:1878

[7] Cheevaporn V, Duangkaew K, Tangkrock-olan N. Environmental occurrence of organochlorine in the East Coast of Thailand 2005. Journal of Health Science 51, 80-88

[8] Davis R.P, Thomas M.R, Garthwaite D.G and Bowen H.M (1992). Pesticide Usage Survey Report 108: Arable Farm Crops in Great Britain, MAFF, London

[9] Doong R.A, Sun Y.C, Liao P.L, Peng C.K, Wu S.C. The distribution and fate of organo-chlorine pesticides residues in sediments from the selected rivers in Taiwan 2002. Chemosphere 48, 237-246

[10] Fabricius K.E. Effects of terrestrial runoff on the ecology on corals and coral reefs: review and synthesis 2005. Marine Pollution Bulletin 50, 125-146

[11] Falco G, Bocio A, Llobet J.M, Domingo J.L, Casas C and Teixido A (2003). Dietary Intake of hexachlorobenzene in Catalonia, Spain. Science of the Total Environment, 322, 1-3

[12] FAO (1999). FAO Yearbook – Production, Food and Agriculture Organization of the United Nations, Rome 43

[13] Forget J, Solomon K.R and Ritter L (1995). A review of selected persistent organic pollutants. International Programme on Chemistry Safety (IPCS) within the framework of the Inter-organization Programme for the sound management of chemicals (IOMC)

[14] Fu J, Mai B, Sheng G, Zhang G, Wang X, Peng P, Xiao X, Ran R, Cheng F, Peng X, Wang Z and Tang U.W (2003). Persistent organic pollutants in environment of the Pearl River Delta, China: An overview. Chemosphere 52, 1411-1422

[15] Golfinopoulos S.K, Anastacia N.D, Lekkas T and Vagi M.C (2003). Organochlorine pesticides in the surface waters of Northern Greece. Chemosphere 50, 507-516

[16] Hong, H., Xu, L., Zhang, J., Chen, J., Woog, Y., Wen, T. Environmental fate and Chemistry of organic pollutants in sediment of Xiamen and Victoria harbors 1995. Marine Pollution Bulletin 31, 229–236

[17] Hung D.Q and Thiemann Q (2003). Contamination by selected chlorinated pesticides in surface waters in Hanoi, Vietnam. Chemosphere 47: 357-367

[18] Imo S.T, Sheikh M.A, Hirosawa E, Oomori T, Tamaki F. Contamination by organo-chlorine pesticides from the selected rivers 2007. International Journal of Environmental Science and Technology (4), 1: 9-17

[19] Iwata H, Tanabe S, Ueda K, Tatsukawa R.P. Persistent organochlorine residues in air, sediments and soils from the lake Baikal Region Russia 1995. Environmental Science and Technology 29, 272-301

[20] Iwata H, Tanabe S, Ueda K, Tatsukawa R.P. Persistent organochlorine residues in air, sediments and soils from the lake Baikal Region Russia 1995. Environmental Science and Technology 29, 272-301

[21] Kennicutt II, M.C., Wade, T.L., Presley, B.J., Requejo, A.G., Brooks, J.M., Denoux, G.J. Sediment contaminants in Casco Bay, Marine: inventories, sources, and Potential for biological impact 1994. Environmental Science and Technology 28, 1-15.

[22] Laabs V, Amelung W, Pinto A.A, Wantzen M, C.J da Silva and Zech W. Pesticides in surface waters, sediments and rainfall of the Northeastern Pantanal Basin, Brazil 2002. Journal of Environmental and Quality 31: 1636-1648

[23] Lee, K.T., Tanabe, S., Koh, C.H. Distribution of organochlorine pesticides in sediments from Kyeonggi Bay and nearby areas, Korea 2001. Environ. Pollut. 114, 207–213.

[24] Miliadis G.E (1993). Gas Chromatographic determination of pesticides in natural waters of Greece. Bulletin Environmental Contamination and Toxicology 50, 247

[25] Monirith I, Ueno D, Takahashi S, Nakata H, Sudaryanto A, Subramanian A, Karuppiah S, Ismail A, Muchtar M, Zheng J, Richardson B.J, Prudente M, Hue D.N, Tana T.S, Tkalin A.V, Tanabe S. Asia Pacific mussel watch: Monitoring contamination of persistent organochlorine compounds in coastal waters of Asian countries 2003. Marine Pollution Bulletin 46, 281-300

[26] Mwevura H, Othman C.O and Mhehe G.L (2002). Organochlorine pesticide residues in edible biota from the Coastal Area of Dar Es Salaam City. Journal of Marine Science 1: 91-96

[27] Nakai K, Nakamura T, Suzuki K, Oka T, Okamuri J, Sugawara N, Saitoh Y, Onba T, Kameo S and Satoh H (2004). Organochlorine pesticides residues in human break milk and placenta in Tohoku, Japan. Organohalogen compounds, 66:2567-2572

[28] Quyen P.B, Dan D.N and Nguyen V.S (1995). Environmental pollution in Vietnam: Analytical estimation and environmental priorities. Trends in Analytical Chemistry, 8:383-388

[29] Sakar A, Nagarajan R, Chaphadkars S, Pal S, Singbal S.Y.S. Contamination of organo-chlorine pesticides in sediments from the Arabian Sea along the west coast of India 1997. Water Resources 31, 195-200

[30] Sheikh M.A, Oomori T, Odo Y. PCBs in sub-tropical marine environment: A case study in seawaters collected around Ryukyus Islands, Japan 2002. Bulletin of Faculty of Science, University of the Ryukyus 74, 65-72

[31] Statistisches Bundesamt (1993). Statistiches Jahrbuch fuer die Bunderrepublik Deutsch-land, Wiesbadener Graphische Betriebe, Wiesbaden (in German)

[32] Strandberg, B., van Bavel, B., Bergqvist, P.A., Broman, D., Ishaq, R., Naf, C., Pettersen, H., Rappe, C. Occurrence, sedimentation, and spatial variations of organochlorine contaminants in settling particulate matter and sediments in the northern Part of the Baltic 1998. Sea. Environ. Sci. Technol. 32, 1754–1759

[33] Tanabe S, Iwata H, Tatsukawa R. Global contamination of persistent organochlorine and their ecotoxicological impact on marine mammals 1994. Science of the Total Environment 154, 163-177

[34] Tashiro Y, Ogura G, Kunisue T, Tanabe S. Persistent organochlorine accumulated in Mongooses (Herpestes javanicus) from Yambaru Area, Okinawa, Japan. Abstract submitted to the 21st Pacific Science Congress, Okinawa 2007. 353

[35] Tashiro Y, Takahira K, Osada H, Fujii H, Tokuyama A. Distribution of polychlorinated biphenyls (PCBs), lead and cadmium in Manko Tidal Flat, 2003, Okinawa. Limnology 5: 177-183

[36] Turnbull A (1998). Chlorinated Pesticides, 113-135

[37] US EPA. Methods for collection, storage and manipulation of sediments for chemical and toxicological analyse 2012

[38] Vagi M.C, Petsas A.S, Karamanoli M.K, Kostopoulou M.N. Determination of organo-chlorine pesticides in marine sediments samples using ultrasonic solvent extraction followed by GC-ECD. Proceedings of the 9th International Conference on Environmental Science and Technology Greece 2005.

[39] Walker K. Vallero D.A, Lewis R.G. Factors influencing the distribution of lindane and other hexachlorocyclohexanes in the environment 1999. Environmental Science and Technology 33, 4373-4378

[40] World Health Organization (2004). Recommended Classification of Pesticides by hazard and Guidelines to Classification

[41] Wu, Y., Zhang, J., Zhou, Q. Persistent organochlorine residues in sediments from Chinese River/Estuary Systems 1999. Environ. Pollut. 105, 143–150

[42] Zhang Z, Hong H, Zhou J, Yu G, Chen W, Wang Z. Transport and fate of organochlorine pesticides in the river water Wuchuan, southeast China 2002. Environmental Monitoring 4, 435-441

[43] Zhou R, Zhu L, Yang K, Chen Y. Distribution of organochlorine pesticide in surface water and sediments from Qiantang River, East China 2006. Journal of Hazardous Materials 137(1): 67-75

Chapter 5

Impact of Systemic Insecticides on Organisms and Ecosystems

Francisco Sánchez-Bayo, Henk A. Tennekes and
Koichi Goka

Additional information is available at the end of the chapter

1. Introduction

Systemic insecticides were first developed in the 1950s, with the introduction of soluble organophosphorus (OP) compounds such as dimethoate, demeton-S-methyl, mevinphos and phorate. They were valuable in controlling sucking pests and burrowing larvae in many crops, their main advantage being their translocation to all tissues of the treated plant. Systemic carbamates followed in the 1960s with aldicarb and carbofuran. Since then, both insecticidal classes comprise a large number of broad-spectrum insecticides used in agriculture all over the world. Nowadays, OPs are the most common pesticides used in tropical, developing countries such as the Philippines and Vietnam, where 22 and 17% of the respective agrochemicals are 'extremely hazardous' [126], i.e. classified as WHO class I. Systemic insect growth regulators were developed during the 1980-90s, and comprise only a handful of compounds, which are more selective than their predecessors. Since 1990 onwards, cartap, fipronil and neonicotinoids are replacing the old hazardous chemicals in most developed and developing countries alike [137].

Through seed coatings and granular applications, systemic insecticides pose minimal risk of pesticide drift or worker exposure in agricultural, nurseries and urban settings. Neonicotinoids and fipronil are also preferred because they appear to be less toxic to fish and terrestrial vertebrates. Initially proposed as environmentally friendly agrochemicals [129], their use in Integrated Pest Management (IPM) programs has been questioned by recent research that shows their negative impact on predatory and parasitic agents [221, 258, 299]. New formulations have been developed to optimize the bioavailability of neonicotinoids, as well as combined formulations with pyrethroids and other insecticides with the aim of broadening the insecticidal spectrum and avoid resistance by pests [83]. Indeed, as with any other chemical used in pest control, resistance to imidacloprid by whitefly (*Bemisia tabaci*), cotton aphids (*Aphis gossypii*) and other pests is rendering ineffective this and other neonicotinoids such as acetamiprid, thiacloprid and nitenpyram [247, 269].

This chapter examines the negative impacts that systemic insecticides have on organisms, populations and ecosystems. The efficacy of these products in controlling the target pests is assumed and not dealt with here – only the effects on non-target organisms and communities are considered.

2. Exposure to systemic insecticides

Unlike typical contact insecticides, that are usually taken up through the arthropod's cuticle or skin of animals, systemic insecticides get into the organisms mainly through feeding on the treated plants or contaminated soil. Thus, monocrotophos and imidacloprid are more lethal to honey bees (*Apis mellifera*) through feeding than contact exposure [143]. Residual or contact exposure affects also some pests and non-target species alike.

Systemic insecticides are applied directly to the crop soil and seedlings in glasshouses using flowable solutions or granules, and often as seed-dressings, with foliar applications and drenching being less common. Being quite water soluble (Table 1), these insecticides are readily taken up by the plant roots or incorporated into the tissues of the growing plants as they develop, so the pests that come to eat them ingest a lethal dose and die. Sucking insects in particular are fatally exposed to systemic insecticides, as sap carries the most concentrated fraction of the poisonous chemical for a few weeks [124], whereas leaf-eating species such as citrus thrips and red mites may not be affected [30]. Systemic insecticides contaminate all plant tissues, from the roots to leaves and flowers, where active residues can be found up to 45-90 days [175, 187], lasting as long as in soil. Thus, pollen and nectar of the flowers get contaminated [33], and residues of imidacloprid and aldicarb have been found at levels above 1 mg/kg in the United States [200]. Guttation drops, in particular, can be contaminated with residues as high as 100-345 mg/L of neonicotinoids during 10-15 days following application [272]. Because these insecticides are incorporated in the flesh of fruits, the highly poisonous aldicarb is prohibited in edible crops such as watermelons, as it has caused human poisoning [106].

As with all poisonous chemicals spread in the environment, not only the target insect pests get affected: any other organism that feeds on the treated plants receives a dose as well, and may die or suffer sublethal effects. For example, uptake of aldicarb by plants and worms results in contamination of the vertebrate fauna up to 90 days after application [41], and honey bees may collect pollen contaminated with neonicotinoids to feed their larvae, which are thus poisoned and die [125]. Newly emerged worker bees are most susceptible to insecticides, followed by foraging workers, while nursery workers are the least susceptible within 72 h of treatment [80]. Insects and mites can negatively be affected by systemic insecticides whenever they feed on:

1. pollen, nectar, plant tissue, sap or guttation drops contaminated with the active ingredient (primary poisoning);

2. prey or hosts that have consumed leaves contaminated with the active ingredient (secondary poisoning).

Parasitoids may be indirectly affected because foliar, drench or granular applications may decrease host population to levels that are not enough to sustain them. Furthermore, host quality may be unacceptable for egg laying by parasitoid females [54]. Small insectivorous animals (e.g. amphibians, reptiles, birds, shrews and bats) can also suffer from primary poisoning if the residual insecticide or its metabolites in the prey are still active. It should be noticed that some metabolites of imidacloprid, thiamethoxam, fipronil and 50% of carbamates are as toxic as the parent compounds [29]. Thus, two species of predatory miridbugs were negatively affected by residues and metabolites of fipronil applied to rice crops [159]. However, since systemic insecticides do not bioaccumulate in organisms, there is little risk of secondary poisoning through the food chain.

Apart from feeding, direct contact exposure may also occur when the systemic insecticides are sprayed on foliage. In these cases, using a silicone adjuvant (Sylgard 309) reduces the contact exposure of honey bees to carbofuran, methomyl and imidacloprid, but increases it for fipronil [184]. In general the susceptibility of bees to a range of insecticides is: wild bees > honey bee > bumble bee [185]. In reality a combination of both contact and feeding exposure occurs, which is more deadly than either route of exposure alone [152, 218].

In soil, residues of acephate and methomyl account for most of the cholinesterase inhibition activity found in mixtures of insecticides [233]. Fortunately, repeated applications of these insecticides induces microbial adaptation, which degrade the active compounds faster over time [250]. Degradation of carbamates and OPs in tropical soils or vegetation is also faster than on temperate regions, due mainly to microbial activity [46]. Some neonicotinoids are degraded by soil microbes [172], and the yeast *Rhodotorula mucilaginosa* can degrade acetamiprid but none of the other neonicotinoids [63], which are quite persistent in this media (Table 2).

Chemical	Group	Vapour Pressure (mPa, 25°C)	Solubilityin water (mg/L)	Log Kow#	GUS index*	Leaching potential
aldicarb	C	3.87	4930	1.15	2.52	moderate
bendiocarb	C	4.6	280	1.72	0.77	low
butocarboxim	C	10.6	35000	1.1	1.32	low
butoxycarboxim	C	0.266	209000	-0.81	**4.87**	high
carbofuran	C	0.08	322	1.8	**3.02**	high
ethiofencarb	C	0.5	1900	2.04	**3.58**	high
methomyl	C	0.72	55000	0.09	2.20	marginal
oxamyl	C	0.051	148100	-0.44	2.36	moderate
pirimicarb	C	0.43	3100	1.7	2.73	moderate

Chemical	Group	Vapour Pressure (mPa, 25°C)	Solubility in water (mg/L)	Log Kow#	GUS index*	Leaching potential
thiodicarb	C	5.7	22.2	1.62	-0.24	low
thiofanox	C	22.6	5200	2.16	1.67	low
triazamate	C	0.13	433	2.59	-0.9	low
cartap	D	1.0×10^{-10}	200000	-0.95	-	high
halofenozide	IGR	<0.013	12.3	3.34	**3.75**	high
hexaflumuron	IGR	0.059	0.027	**5.68**	-0.03	unlikely to leach
novaluron	IGR	0.016	0.003	**4.3**	0.03	low
teflubenzuron	IGR	0.000013	0.01	**4.3**	-0.82	low
acetamiprid	N	0.000173	2950	0.8	0.94	low
clothianidin	N	2.8×10^{-8}	340	0.905	**4.91**	high
dinotefuran	N	0.0017	39830	-0.549	**4.95**	high
imidacloprid	N	0.0000004	610	0.57	**3.76**	high
nitenpyram	N	0.0011	590000	-0.66	2.01	moderate
thiacloprid	N	0.0000003	184	1.26	1.44	unlikely to leach
thiamethoxam	N	0.0000066	4100	-0.13	**3.82**	high
acephate	OP	0.226	790000	-0.85	1.14	low
demeton-S-methyl	OP	40	22000	1.32	0.88	low
dicrotophos	OP	9.3	1000000	-0.5	**3.08**	high
dimethoate	OP	0.25	39800	0.704	1.06	low
disulfoton	OP	7.2	25	3.95	1.29	low
fenamiphos	OP	0.12	345	3.3	-0.11	low
fosthiazate	OP	0.56	9000	1.68	2.48	moderate
heptenophos	OP	65	2200	2.32	0.26	low
methamidophos	OP	2.3	200000	-0.79	2.18	moderate
mevinphos	OP	17	600000	0.127	0.19	low
monocrotophos	OP	0.29	818000	-0.22	2.3	moderate
omethoate	OP	3.3	10000	-0.74	2.73	moderate
oxydemeton-methyl	OP	2.0	1200000	-0.74	0.0	low
phorate	OP	112	50	3.86	1.4	low
phosphamidon	OP	2.93	1000000	0.79	2.39	moderate
thiometon	OP	39.9	200	3.15	0.37	low
vamidothion	OP	1.0×10^{-10}	4000000	-4.21	0.55	low
fipronil	PP	0.002	3.78	3.75	2.45	moderate

Table 1. Physicochemical properties of systemic insecticides. C = carbamates; D = dithiol; IGR = Insect growth regulator; N = neonicotinoid; OP = organophosphate; PP = phenylpyrazole
Partition coefficients between n-octanol and water (Kow) indicate bioaccumulation potential when Log Kow > 4.
*The Groundwater Ubiquity Score (GUS) is calculated using soil half-life (DT50) and organic-carbon sorption constant (Koc) as follows: GUS = log(DT50) x (4-log Koc). A compound is likely to leach if GUS > 2.8 and unlikely to leach when GUS < 1.8; other values in between indicate that leaching potential is marginal.

Chemical	Group	Water		Field	
		Photolysis (pH 7)	Hydrolysis (pH 5-7)	Water-sediment	Soil (range)
aldicarb	C	8	**189**	6	10 (1-60)
bendiocarb	C	13	25	2	4 (3-20)
butocarboxim	C	Sta ble	stable	-	4 (1-8)
butoxycarboxim	C	Stable	18 (510-16)	-	42
carbofuran	C	71	37 (46-0.1)	9.7	14 (1-60)
ethiofencarb	C	-	16	52	37 (34-131)
methomyl	C	Stable	stable	4	7 (5-30)
oxamyl	C	7	8	<1	11
pirimicarb	C	6	stable	**195**	9 (5-13)
thiodicarb	C	9	30 (69-0.3)	<1	18 (1-45)
thiofanox	C	1	30	-	4 (2-6)
triazamate	C	301	2	<1	<1
cartap	D	-	-	-	3
halofenozide	IGR	10	stable	-	**219** (60-219)
hexaflumuron	IGR	6	stable	-	**170**
novaluron	IGR	Stable	stable	18	97 (33-160)
teflubenzuron	IGR	10	stable	16	14 (9-16)
acetamiprid	N	34	**420**[a]	-	3 (2-20)
clothianidin	N	0.1	14 [a]	56	**545** (13-1386)
dinotefuran	N	0.2	stable	-	82 (50-100)
imidacloprid	N	0.2	~ **365** [a]	**129**	**191** (104-228)
nitenpyram	N	NA	2.9 [a]	-	8
thiacloprid	N	stable	stable	28	16 (9-27)
thiamethoxam	N	2.7	11.5 [a]	40	50 (7-72)
acephate	OP	2	50	-	3
demeton-S-methyl	OP	-	56 (63-8)	-	2.7
dicrotophos	OP	-	-	-	28
dimethoate	OP	175	68 (156-4)	15	7 (5-10)
disulfoton	OP	4	**300**	15	30
fenamiphos	OP	<1	**304**	60	2 (1-50)
fosthiazate	OP	Stable	**104** (178-3)	51	13 (9-17)
heptenophos	OP	-	13	7	1
methamidophos	OP	90	5	24	4 (2-6)
mevinphos	OP	27	17	21	1 (1-12)
monocrotophos	OP	26	**134**	-	30 (1-35)
omethoate	OP	Stable	17	5	14
oxydemeton-methyl	OP	222	73 (96-41)	3	5
phorate	OP	1	3	-	63 (14-90)

Chemical	Group	Water		Field	
phosphamidon	OP	-	36 (60-12)	13	12 (9-17)
thiometon	OP	-	22	-	2 (2-7)
vamidothion	OP	-	**119**	7	1 (<1-2)
fipronil	PP	0.33	stable	68	65 (6-135)

Table 2. Degradation of systemic insecticides expressed as half-lives in days. Compounds with half-lives longer than 100 days are considered persistent (Sources: Footprint database & [284]. [a] for pH 9C = carbamates; D = dithiol; IGR = Insect growth regulator; N = neonicotinoid; OP = organophosphate; PP = phenylpyrazole

Aquatic organisms take up easily whatever residues reach the waterbodies, through runoff from treated fields or contaminated groundwater. Some 20% systemic insecticides are prone to leaching, and 45% are mobile in wet soils (Table 1). For example, acephate leaches more easily than methamidophos [305], and so acephate should be restricted or avoided in tropical areas and rice crops [46]. Residues of aldicarb and methomyl in groundwater can have sublethal effects in mammals [215]. Even if residue levels of systemic insecticides in rivers and lakes are usually at ppb levels (μg/L), persistent compounds such as fipronil, neonicotinoids and growth regulators can have chronic effects due to their constant presence throughout several months in the agricultural season [123]. For example, about 1-2% of imidacloprid in treated soil moves into runoff after rainfall events, with the highest concentrations recorded at 0.49 mg/L [12]. Systemic carbamates and OPs do not last long in water because they breakdown through photolysis or hydrolysis in a few days, or are taken up and degraded by aquatic plants [100]. In any case, their presence and frequency of detection in water depends on local usage patterns [39, 171]. The acute toxicity of most systemic compounds is enhanced in aquatic insects and shrimp under saline stress [22, 253].

A characteristic feature of most systemic insecticides –except carbamates– is their increased toxicity with exposure time, which results from a constant or chronic uptake through either feeding or aquatic exposure (Figure 1). Effects are more pronounced some time after the initial application [16], and could last up to eight months [286]. Also, as a result of chronic intoxication, there may not be limiting toxic concentrations (e.g. NOEC or NOEL) in compounds that have irreversible mechanism of toxicity, since any concentration will produce an effect as long as there is sufficient exposure during the life of the organism [274]. This is precisely their main advantage for pest control: any concentration of imidacloprid in the range 0.2-1.6 ml/L can reduce the population of mango hoppers (*Idioscopus* spp.) to zero within three weeks [291]. However, it is also the greatest danger for all non-target species affected, e.g. predators, pollinators and parasitoids. By contrast, contact insecticides act usually in single exposures (e.g. spray droplets, pulse contamination after spraying, etc.) and have the highest effects immediately after application.

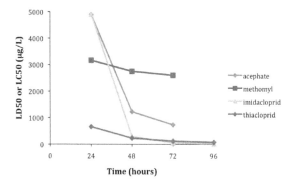

Figure 1. Increasing toxicity of several systemic insecticides with time of exposure. LD50 for acephate to *Episyrphus bateatus* and for methomyl to *Bombus terrestris* [75]; LC50 for imidacloprid to *Cypridopsis vidua* [234] and thiacloprid to *Sympetrum striolatum* [28].

3. Modes of action of systemic insecticides

Before describing their impacts on organisms and ecosystems, a description of the mechanisms of toxicity of systemic insecticides is briefly outlined.

3.1. Acetylcholinesterase inhibitors

Carbamates and organophosphorus compounds are inhibitors of the acetylcholinesterase enzyme (AChE), thus blocking the transmission of the nervous impulse through the neuronal synapses. The binding of carbamates to the enzyme is slowly reversible and temporary, i.e. < 24 h [197], whereas that of alkyl OPs is irreversible. The binding of methyl-OPs does not last as long as that of alkyl-OPs, and this feature is compound specific [182]. Given their mode of action, all these compounds are broad-spectrum insecticides, extremely toxic to most animal taxa, from worms to mammalian vertebrates. Avian species are often more susceptible to these compounds due to relatively low levels of detoxifying enzymes in birds [207, 297]. Thus, recovery of ducklings exposed to a range of carbamate and OP insecticides occurred within eight days after being depressed 25-58% following dosing [91].

3.2. Insecticides acting on nicotinic acetylcholine receptors (nAChR)

Neonicotinoids are derived from nicotine, which is found in the nightshade family of plants (Solanaceae), and particularly in tobacco (*Nicotiana tabacum*). They all are agonists of the nicotinic acetylcholinesterase receptor (nAChR), which mediate fast cholinergic synaptic transmission and play roles in many sensory and cognitive processes in invertebrates. Binding of neonicotinoids to these receptors is irreversible in arthropods [40, 307]. Given that nAChRs are embedded in the membrane at the neuronal synapses, their regeneration seems unlikely because neurons do not grow. The lower affinity of neonicotinoids for mammalian nAChRs has been attributed to the different ionic structure of the vertebrate subtypes [283]. The high toxicity of neonicotinoids to insects and worms is comparable to that of pyrethroids, but aquatic crustaceans, particularly waterfleas, are more tolerant [119, 136].

Cartap is a dithiol pro-insecticide that converts to nereistoxin, a natural toxin found in marine *Nereis* molluscs. Both cartap and nereistoxin are antagonists of the nAChR in insects and other arthropods [164], blocking irreversibly the neuronal functions of these receptors. Unlike neonicotinoids, cartap appears to be very toxic to fish and amphibians [235].

3.3. GABA-R antagonists (fipronil)

Fipronil is a phenylpyrazole antagonist of the γ-aminobutyric acid (GABA)-gated chloride channel, binding irreversibly to this receptor and impeding the nervous transmission [56]. Its mode of action, therefore, appears to be identical to that of cyclodiene organochlorins (e.g. endosulfan), but fipronil is mostly systemic whereas all cyclodienes are insecticides with contact activity. Interestingly, while aquatic organisms (e.g. cladocerans, fish) are quite tolerant of fipronil, vertebrates are more susceptible to this compound than to the old organochlorins [235].

3.4. Insect growth regulators (IGR)

Hexaflumuron, novaluron and teflubenzuron are the only systemic benzoylureas in the market. They are chitin inhibitors, blocking the biosynthesis of this essential component of the arthropod's exoskeleton. As a consequence, insects and other arthropods cannot moult and die during their development. Since their mode of action is restricted to arthropods, benzoylureas are not very toxic to any other animal taxa, e.g. molluscs, vertebrates, etc. [235].

Halofenozide is the only systemic compound among the hydrazines, a group of chemicals that mimic the steroidal hormone ecdysone, which promotes moulting in arthropods [71]. The premature moulting in larvae of some insect taxa, particularly in Lepidoptera, prevents them from reaching the adult stage. Toxicity of halofenozide is selective to insects only.

4. Effects on organisms and ecosystems

4.1. Direct effects on organisms

Mortality of non-target organisms exposed to insecticides is mostly due to acute toxicity, particularly in the case of carbamates. However, with systemic compounds there are many observations of long-term suppression of populations that suggest a chronic lethal impact over time. The latter impacts are likely due to persistence of residual activity in the soil, foliage or water in the case of reversible toxicants (i.e. carbamates), or to irreversible and persistent binding in other cases. (note: all application rates and concentrations here refer to the active ingredient).

4.1.1. Acetylcholinesterase inhibitors

These compounds can have serious impacts on soil organisms of various taxa. Aldicarb and phorate applied to a cotton crop soil at 0.5 and 1 kg/ha, respectively, eliminated or reduced significantly non-target mesofauna, including mites and springtails. Populations of the latter taxa were reduced for more than 60 days (phorate) and 114 days (aldicarb) [17, 225], with the highest effects peaking after 18 days [16]. Granular applications of phorate (250 mg/kg dry soil) killed almost all earthworms, Collembola, Acarina, free-living saprophytic and parasitic nematodes and Protozoa, with populations of Collembola recovering only when residues went below 2 mg/kg [300]. After a single aldicarb application to soil at 2.5 g/m², Gamasina predatory mites went to extinction within a year [148]. Bendiocarb impacts on predaceous arthropods and oribatid mites were less severe and temporary compared to the impacts of non-systemic OPs, but increased trap catches of ants two weeks after application [55], possibly as a result of a longer-term effect. Many soil arthropods, in particular mites and springtails, were the most affected by dimethoate –and its metabolite omethoate– residues in soil after sprays of 1-2 ml/L in the farms of the Zendan valley, Yemen [4]. Similar observations were made when dimethoate was sprayed on vegetation of arable fields [85] or in soil microcosms [180]; the springtail populations recovered but attained lower densities a year later, while their dominance structure had changed. However, dimethoate or phosphamidon applied in mustard fields produced only a temporary decline, compared to the long-lasting effect of monocrotophos [141]. Collembola populations do not seem to be affected by pirimicarb applications on cereal crops [95].

Earthworm populations were affected initially after application of phorate and carbofuran to turfgrass, but not thiofanox, and their numbers recovered subsequently [53]. Reduction of earthworm populations by bendiocarb was the highest (99% in one week) among 17 insecticides applied at label rates on turfgrass, with significant effects lasting up to 20 weeks [216]. Juveniles and species living in the surface layers or coming to the soil surface to feed (e.g., *Lumbricus terrestris*) are most affected, since a high degree of exposure is usually found in the first 2.5 cm of soil [288]. However, systemic carbamates can be selective to plant-parasitic nematodes without affecting fungal or microbial communities [296]. Thus, cholinesterase inhibitors do not have significant impacts on bacteria, fungi and protozoa in soil [133], and consequently do not alter the soil biochemical processes [79]. Nevertheless, a combined dimethoate-carbofuran application reduced active hyphal lengths and the number of active bacteria in a treated forest soil [58].

Populations of beneficial predators can be decimated initially as much as the target pests, but they usually recover quickly. For example, thiodicarb or its degradation product, methomyl, applied at 0.5 kg/ha on soybean crops, significantly reduced populations of the predatory bugs *Tropiconabis capsiformis* and *Nabis roseipennis* within two days after treatment only [25]. Demeton-S-methyl reduced populations of predatory insects on strawberry patches, whereas pirimicarb and heptenophos had no significant effect on spiders, staphylinids and anthocorids, or on hymenopteran parasitoids [76]. While populations of web spiders and carabid beetles are severely reduced by dimethoate applied to cabbage fields and cereal crops [144], pirimicarb does not seem to have much impact on these taxa [97, 195], affecting mainly aphids [131]. Pirimicarb on wheat crops does not impact on ladybirds, but larvae of *Episyrphus balteatus* are affected [135]. By contrast, longer impacts have been observed with acephate applied at 0.5 kg/ha on rice paddies, which reduced populations of predatory bugs (*Cyrtorrhinus lividipennis* and *Paederus fuscipes*) for at least 10 days [155]. Similar rates of acephate on rice and soybean crops reduced spiders populations for three weeks, but they recovered afterwards [181]. In addition, acephate is deadly to three species of whitefly parasitoid species [267].

Direct mortality of bumble bees (*Bombus terrestris*) in short exposures to dimethoate is much higher than for heptenophos or ethiofencarb [132]. However, what matters most is the chronic toxicity to the entire bee colony not just the workers. For example, methamidophos contaminated syrup (2 mg/L) produced significant losses of eggs and larvae of honey bees without any appreciable loss of workers after one week of exposure; the colonies would recover completely within 13 weeks if the insecticide was applied only once [301], indicating a long-term impact on the colony. Similarly, the mortality of non-target adult chrysomelid beetles (*Gastrophysa polygoni*) after foliar treatment with dimethoate on the host plants was low (1.9-7.6%), but because this insecticide was most toxic to the egg stage, the overall beetle population decreased over time due to hatching failure [146].

Primary poisoning of birds and mammals by ingestion of OP and carbamate granules or coated seeds is still a problem despite the many attempts to reduce these impacts [189, 190]. For example, mortality of birds that ingested granules of carbofuran in a corn field was extensive, affecting waterfowl, small songbirds and mice within 24 hours. Residues up to 17 mg/kg body weight (b.w.) were found in the dead animals [19]. The granular formulation of this carbamate was banned in the mid-1990s by the US EPA after numerous cases of direct poisoning by animals; however, the liquid formulation applied to alfalfa and corn is just as deadly to bees, because this systemic insecticide is present in the pollen of those plants [208]. Phosphamidon sprayed at 1 kg/ha to larch forests in Switzerland caused many bird deaths [243]; large bird mortality was also observed in Canadian spruce forests sprayed with phosphamidon (0.55 kg/ha), particularly among insectivorous warblers. There was good evidence that birds picked up the insecticide from sprayed foliage within a few hours of application [94]. Carbofuran and phosphamidon were the most common pesticides implicated in deaths of wild birds in Korea between 1998-2002 [157], and ducklings died in large numbers when phorate was applied to South Dakota wetlands [73]. Usually birds die when their brain AChE depression is over 75% [92, 114]. Thus, 11 out of 15 blue jays (*Cyanocitta cristata*) which had depression levels ranging 32-72% after disulfoton was sprayed to pecan groves would die [302], but their carcasses would probably not be found. In orchards sprayed with methomyl, oxamyl or dimethoate, the daily survival rates for nests of Pennsylvania mourning dove (*Zenaida macroura*) and American robin (*Turdus migratorius*) were significantly lower than in non-treated orchards, and the species diversity was also lower. Repeated applications of these and other insecticides reduced the reproductive success of doves and robins and may have lowered avian species diversity [93].

Secondary poisoning with bendiocarb was attributed to 22 birds that had depressed AChE activity after eating contaminated mole crickets and other soil organisms on the applied turfgrass [224]. Several species of raptors were killed or debilitated after consuming water-fowl contaminated with phorate – the fowl had ingested granules of the insecticide that were applied to potato fields a few months earlier [84]. Equally, ladybugs (*Hippodamia unde-cimnotata*) fed upon *Aphis fabae*, which were reared on bean plants treated with carbofuran, experienced a 67% population reduction due to secondary poisoning [206]. Pirimicarb caused 30-40% mortality of Tasmanian brown lacewing (*Micromus tasmaniae*) larvae when feeding on contaminated 1st instar lettuce aphid (*Nasonovia ribisnigri*) for three days [298].

Impacts on aquatic organisms usually do not last more than a month. For example, thiodi-carb applied at 0.25-1.0 kg/ha had severe impacts on copepods, mayflies and chironomids in experimental ponds for three weeks, but not so much on aquatic beetle's larvae; eventually there was recovery of all populations [7]. Pirimicarb can be lethal to common frog (*Rana tem-poraria*) tadpoles, but does not appear to have chronic effects [139]. However, vamidothion and acephate are most lethal to non-target organisms in rice crops, and are not recommend-ed in IPM programs [153]. Carbofuran and phorate are very toxic to aquatic invertebrates [140], particularly amphipods and chironomids but not so much to snails, leeches or ostra-cods [72, 249]. Small negative effects in zooplankton communities (cladocerans copepods and rotifers) were observed in rice paddies treated with carbofuran at recommended appli-cation rates, but fish were not affected [107]. Carbofuran should not be used in rice paddies, whether in foliar or granular formulations: not only induces resurgence of the brown plan-thoppers (*Nilaparvata lugens*) [122], but it is also more toxic to the freshwater flagellate *Eugle-na gracilis* than the non-systemic malation [15]. It reduces populations of coccinellid beetles, carabid beetles, dragonfly and damselfly nymphs, but does not impact much on spiders [255]. However, it appears that carbofuran at 0.2% per ha can double the densities of *Steno-cypris major* ostracods in rice paddies, whereas other insecticides had negative effects on this species [168]. Repeated applications of carbofuran can also have a significant stimulation of the rhizosphere associated nitrogenase activity, with populations of nitrogen-fixing *Azospir-illum* sp., *Azotobacter* sp. and anaerobic nitrogen-fixing bacteria increasing progressively up to the third application of this insecticide [142].

4.1.2. Insecticides acting on nAChR

Direct toxicity of cartap to fish species is not as high as that of other neurotoxic insecticides, with 3-h LC50s between 0.02 and 6.8 mg/L [161, 308]. However, cartap affects negatively several species of Hymenoptera and aphid parasitoids used to control a number of crop pests [14, 77, 147, 270]. This insecticide also inhibits hatching of eggs of the nematode *Aga-mermis unka*, a parasite of the rice pest *Nilaparvata lugens* [50], and reduces significantly the populations of ladybugs and other predatory insects in cotton crops when applied at the rec-ommended rates, i.e. 20 g/ha [109, 169]. In rice paddies, cartap hydrochloride reduced popu-lations of coccinellid beetles, carabid beetles, dragonflies and damselflies by 20-50% [255]. Pollinators such as honey bees and bumble bees can also be seriously reduced in numbers when feeding on crops treated with cartap hydrochloride, which is included among the most toxic insecticides to bees after neonicotinoids and pyrethroids [179, 278]. For all its neg-ative impacts on parasitoids and predatory insects it is hard to understand why cartap was the third most common insecticide (19% of all applications) used in IPM programs in Vietnam a decade ago [31], and is still among the most widely used in rice farms in China [308].

Cumulative toxicity of neonicotinoids over time of exposure results in long-term pest control compared to the impact of cholinesterase inhibitor insecticides. For example, soil treated with clothianidin at 0.05-0.15% caused increasing mortality in several species of wireworms (Coleoptera: Elateridae), reaching 30-65% after 70 days, whereas chlorpyrifos at 0.15% produced 35% mortality within 30 days but no more afterwards [292]. Soil application of imidacloprid did not eliminate rapidly Asian citrus psyllid (*Diaphorina citri*) and leafminer (*Phyllocnistis citrella*) populations, but resulted in chronic residues in leaf tissue and long-term suppression of both pests [245]. Also, soil applications of neonicotinoids are very effective in controlling soil grubs and berry moths (*Paralobesia viteana*) in vineyards provided there is no irrigation or rain that washes off the insecticide [289]. For the same reason, however, the impact of neonicotinoids on non-target organisms is long-lasting. For example, repeated corn-seed treatement with imidacloprid caused a significant reduction in species richness of rove beetles in three years, even though the abundance of the main species was not affected [88]. In addition to long-term toxicity, acute toxicity of acetamiprid, imidacloprid and thiomethoxam to planthopper and aphid species is similar to that of synthetic pyrethroids, and higher than that of endosulfan or acetylcholinesterase inhibitors [219, 246]. Thus, combinations of pyrethroid-neonicotinoid have been hailed as the panacea for most pest problems as it suppresses all insect resistance [70]. Mixtures of imidacloprid and thiacloprid had additive effects on the toxicity to the nematode *Caenorhabditis elegans* but not on the earthworm *Eisenia fetida* [108].

Acute toxicity of imidacloprid, thiamethoxam, clothianidin, dinotefuran and nitenpyram to honey bees is higher than that of pyrethroids, while toxicity of acetamiprid and thiacloprid is increased by synergism with ergosterol-inhibiting fungicides [134, 242] and antibiotics [116]. Thus, neonicotinoids can pose a high risk to honey bees, bumble bees [176, 263] and wasps [90]. Bees can be killed immediately by direct contact with neonicotinoid droplets ejected from seed drilling machines. Thus, numerous worker bees were killed when seed was coated with clothianidin during drilling of corn in the Upper Rhine Valley (Germany) in spring 2008 [102]. The same problem happened in Italy with thiamethoxam, imidacloprid and clothianidin [105, 285], leading to the banning of this application method on sunflower, canola and corn during 2008-09 [20]. However, most of the time bee colonies are intoxicated by feeding on contaminated pollen and nectar [9, 228]. It has been observed that bee foraging was notably reduced when Indian mustard was treated with 178 mg/ha imidacloprid [10]. Imidacloprid residues in sunflowers are below the no-adverse-effect concentration to honey bees of 20 µg/kg at 48-h [241], with surveys in France showing residue levels in pollen from treated crops in the range 0.1-10 µg/kg and average in nectar of 1.9 µg/kg [33]. However, bees feeding on such contaminated pollen or nectar will reach first sublethal and later lethal levels, with 50% mortality occurring within 1-2 weeks [228, 266]. Such data was disputed [89, 240] as it was in conflict with some long-term field observations of honey bees feeding on sunflowers grown from imidacloprid-treated seeds at 0.24 mg/seed [256]. However, recent evidence suggest that chronic lethality by imidacloprid is implicated in the colony collapse disorder (CCD) that affects honey bees [174]. Based on the fast degradation of imidacloprid in bees (4-5 hours), it is assumed that honey bees which consume higher amounts of imidacloprid die already outside of the hive, before the colony's demise and before samples are taken, though residues of imidacloprid in bees at 5-8 µg/kg have been found in some cases [111]. Clothianidin residues of 6 µg/kg in pollen from canola fields reduced the number of bumble bee (*Bombus impatiens*) workers slightly (~20%) [96], but exposure to clothianidin-treated canola for three weeks appeared not to have affected honey bee colonies in Canada [61]. Thiamethoxam applied to tomatoes (~150 g/ha) through irrigation water does not have impacts on bumble bees (*Bombus terrestris*) [244], whereas pollen contaminated with this insecticide causes high mortality and homing failure [125].

Negative impacts of neonicotinoids on non-target soil arthropods are well documented. A single imidacloprid application to soil reduced the abundance of soil mesofauna as well as predation on eggs of Japanese beetle (*Popillia japonica*) by 28-76%, with impacts lasting four weeks. The same level of impact was observed with single applications of clothianidin, dinotefuran and thiamethoxam, so the intended pest control at the time of beetle oviposition runs into conflict with unintended effects – disruption of egg predation by non-target predators [210]. Among several insecticides applied to home lawns, only imidacloprid suppressed the abundance of Collembola, Thysanoptera and Coleoptera adults, non-oribatid mites, Hymenoptera, Hemiptera, Coleoptera larvae or Diptera taxonomic groups by 54-62% [209]. Imidacloprid applied to the root of eggplants (10 mg/plant) greatly reduced most arthropod communities and the species diversity during the first month. Small amounts of soil residues that moved into the surrounding pasture affected also some species; however, non-target ground arthropods both inside and outside the crop showed significant impacts only in the two weeks after planting [238], probably due to compensatory immigration from nearby grounds.

Foliar applications of thiamethoxam and imidacloprid on soybean crops are preferred to seed treatments, as neonicotinoids appear to have lesser impacts on non-target communities than pyrethroids [204]. However, a foliar application of thiacloprid (0.2 kg/ha) to apple trees reduced the population of earwigs (*Forficula auricularia*), an important predator of psyllids and woolly apple aphid, by 60% in two weeks, while remaining below 50% after six weeks [294]. Branchlets of hemlock (*Tsuga canadensis*) treated with systemic imidacloprid (1-100 mg/kg) reduced the populations of two non-target predators of the hemlock woolly adelgid (*Adelges tsugae*) and had both lethal and sublethal effects on them [78]. Clothianidin, thiamethoxam and acetamiprid were as damaging to cotton crop predators as other broad-spectrum insecticides and cartap [169]. All neonicotinoids are lethal to the predatory mirid *Pilophorus typicus*, a biological control agent against the whitefly *Bemisia tabaci*, since their residual activity can last for 35 days on the treated plants [201]. The ladybug *Serangium japonicum*, also a predator of the whitefly, is killed in large numbers when exposed to residues of imidacloprid on cotton leaves applied at the recommended rate (40 ppm) or lower; apparently, the predator was not affected when imidacloprid was applied as systemic insecticide [120]. Clothianidin is 35 times more toxic to the predatory green miridbug (*Cyrtorhinus lividipennis* 48-h LC50 = 6 µg/L) than to the main pest of rice (*Nilaparvata lugens* 48-h LC50 = 211 µg/L), thus questioning seriously its application in such crops [221]. Not surprisingly, populations of predatory miridbugs and spiders suffered an initial set back when rice paddies were treated with a mixture of ethiprole+imidacloprid (125 g/ha), and their recovery was slow and never attained the densities of the control plots [154]. Mixtures of ethiprole+imidacloprid and thiamethoxam+l λ-cyhalothrin on rice paddies are also highly toxic to mirid and veliid natural enemies of rice pests, with 100% mortalities recorded in 24 h [159].

Secondary poisoning with neonicotinoids reduces or eliminates eventually all predatory ladybirds in the treated areas, compromising biological control in IPM programs. Indeed, exposure of larval stages of *Adalia bipunctata* to imidacloprid, thiamethoxam, and acetamiprid, and adult stages to imidacloprid and thiamethoxam, significantly reduced all the demographic parameters in comparison with a control –except for the mean generation time–, thus resulting in a reduced coccinellid population; adult exposures produced a significant population delay [162]. Eighty percent of 3rd and 4th instar larvae of the ladybug *Harmonia axyridis* died after feeding for 6 hours on corn seedlings grown from seeds treated with clothianidin, compared to 53% mortality caused by a similar treatment with thiamethoxam; recovery occurred only in 7% of cases [196]. Survival of the ladybird *Coleomegilla maculata* among flower plants treated with imidacloprid at the label rate was reduced by 62% [251], and *Hippodamia undecimnotata* fed upon aphids reared on bean plants treated with imidacloprid, experienced a 52% population reduction [206]. Equally, 96% of Tasmanian brown lacewing (*Micromus tasmaniae*) larvae died after feeding on 1st instar lettuce aphid (*Nasonovia ribisnigri*) for three days. Low doses did not increase mortality but from days 3 to 8, lacewing larvae showed significant evidence of delayed developmental rate into pupae [298]. Grafton-Cardwell and Wu [110] demonstrated that IGRs, neonicotinoid insecticides, and pyrethroid insecticides have a significant, negative impact on vedalia beetles (*Rodolia cardinalis*), which are essential to control scale pests in citrus; neonicotinoids were toxic to vedalia larvae feeding on cottony cushion scale that had ingested these insecticides, and survival of adult beetles was also affected but to a lesser extent than other insecticides.

Recent evidence of the negative impacts of neonicotinoids on parasitoids reinforces that these insecticides are not suitable for IPM [271]. All neonicotinoids are deadly to three whitefly parasitoid species (*Eretmocerus* spp. and *Encarsia formosa*), with mortality of adults usually greater than the pupae [267]. Thiamethoxam appears to be less toxic to whitefly parasitoids compared to imidacloprid [202]. Imidacloprid, thiamethoxam and nitenpyram appeared to be the most toxic to the egg parasitoids *Trichogramma* spp. [231, 299]. For example, the acute toxicity of thiomethoxam and imidacloprid to *Trichogramma chilonis*, an egg parasitoid of leaf folders widely used in cotton IPM, is about 2000 times higher than that of other insecticides used in rice crops in India, such as acephate or endosulfan [220]. Acute toxicity of imidacloprid is more pronounced on Braconidae parasitoids than on *T. chilonis*, whereas thiacloprid only reduced the parasitization on *Microplitis mediator* [192]. Thiacloprid is as toxic to the cabbage aphid *Brevicoryne brassicae* as to its parasitoid (*Diaeretiella rapae*), whereas pirimicarb and cypermethrin are more toxic to the aphid and are, therefore, preferred in IPM [3].

Neonicotinoids pose also risks to aquatic taxa. The synergistic toxicity of imidacloprid+thiacloprid on *Daphnia magna* [173] implies the combined effect of neonicotinoids on aquatic arthropods would be higher than expected, even if *Daphnia* is very tolerant of neonicotinoids [119]. Other contaminants, such as the nonylphenol polyethoxylate (R-11) act also synergistically with imidacloprid [49]. Thiacloprid causes delayed lethal and sublethal effects in aquatic arthropods, which can be observed after 4 to 12 d following exposure to single 24-h pulses [28]. Thus, its 5% hazardous concentration (0.72 µg/L) is one order of magnitude lower than predicted environmental concentrations in water [35]. Also, thiacloprid LC50 for survival of midges (*Chironomus riparius*) is only 1.6 µg/L, and EC50 for emergence 0.54 µg/L [160], so both acute and chronic toxicity reduce the survival and growth of *C. tentans* and *Hyalella azteca* [265]. Acute toxicity of neonicotinoids to red swamp crayfish (*Procambarus clarkii*) is 2-3 orders of magnitude lower than that of pyrethroids [23]; comparative data such as this gives the neonicotinoids an apparent better environmental profile. However, experimental rice mesocosms treated with imidacloprid at label rates (15 kg/ha) eliminated all zooplankton communities for two months, and their recovery did not reach the control population levels four months later. Equally, mayflies, coleoptera larvae and dragonfly nymphs were significantly reduced while residues of imidacloprid in water were above 1 µg/L [117, 237]. Similarly, streams contaminated with a pulse of thiacloprid (0.1-100 µg/L) resulted in long-term (7 months) alteration of the overall invertebrate community structure [27]. However, while aquatic arthropods with low sensitivity to thiacloprid showed only transient effects at 100 µg/L, the most sensitive univoltine species were affected at 0.1 µg/L and did not recover during one year [167].

4.1.3. Fipronil

Fipronil is very efficient in controlling locust outbreaks, but causes more hazards than chlorpyrifos and deltamethrin to non-target insects in the sprayed areas, although it is more selective to specific taxa [214, 252]. Thus, abundance, diversity and activity of termites and ants were all reduced in northern Australia after spraying several areas with fipronil for locust control [262], and 45% of the termite colonies died within 10 months of a spraying operation with fipronil for controlling locusts in Madagascar [214]. Reducing the recommended application rates by seven times (0.6-2 g/ha) still achieves 91% elimination of locusts while having lesser impacts on non-target organisms, comparable to those inflicted by carbamate and OP insecticides [18].

Despite its selectivity, fipronil in maize crops reduced the abundance of arthropod populations of the soil mesofauna more significantly than other systemic insecticides, i.e. carbofuran [59], although springtails are little affected as they avoid feeding on litter contaminated with fipronil and are more tolerant of this insecticide [232]. When applied to citrus orchards, fipronil was among the most detrimental insecticides affecting two *Euseius* spp. of predatory mites [112]. In rice crops, the effectiveness of fipronil in controlling pests was overshadowed by its negative impact on the predatory miridbugs *Cyrtorhinus lividipennis* and *Tytthus parviceps* [159].

Of greater concern is the impact of this systemic chemical on honey bees and wild bee pollinators. With an acute contact LD50 of 3.5 ng/bee [166] and acute oral LD50 of 3.7-6.0 ng/bee [2], fipronil is among the most toxic insecticides to bees ever developed. Even more worrying is the finding that the adjuvant Sylgard, used to reduce the toxicity of most insecticidal products on bees, increases the toxic effects of fipronil [184]. The systemic nature of this chemical implies that chronic feeding of the bees on nectar contaminated with fipronil caused 100% honey bee mortality after 7 days, even if the residue concentration was about 50 times lower than the acute lethal dose [8]. Residues of fipronil in pollen have been measured as 0.3-0.4 ng/g, which are 30-40 times higher than the concentration inducing significant mortality of bees by chronic intoxication [33]. Unlike neonicotinoids, no residues of fipronil have been found in guttation drops [272].

The acute toxicity of fipronil to cladocerans is similar to the toxicity to estuarine copepods, with 48-h LC50 in the range 3.5-15.6 µg/L [47, 259], but the chronic toxicity with time of exposure is what determines the fate of the populations exposed. For example, populations of *Daphnia pulex* went to extinction after exposure to 80 µg/L for 10 days, equivalent to LC75 [259], and 40% of a population of grass shrimps (*Palaemonetes pugio*) died in 28 days after being exposed to fipronil concentrations of 0.35 µg/L in marsh mesocosms, and none of the shrimps survived when exposed to 5 mg/L during the same period [303]. Such impacts on zooplankton are likely to occur in estuaries, where waters have been found to contain 0.2-16 µg/L of fipronil residues [45, 163], even if no apparent effect on amphipods, mussels nor fish has been observed [37, 303]. Fipronil sprays on water surfaces to control mosquito larvae have negative impacts not only on cladocerans but also on chironomid larvae exposed to chronic feeding on contaminated residues [183, 264]. Studies on rice mesocosms have shown that significant population reductions due to fipronil application at the recommended rates (50 g per seedling box) are not restricted to zooplankton and benthic species, but affect most species of aquatic insects. Moreover, fipronil impacts on aquatic arthropods were more pronounced after a second application in the following year [118], indicating persistence of this insecticide in rice paddies. Chronic toxicity over time explains the long-term toxicity of this systemic compound, so it is not surprising that concentrations of 1.3 µg/L in paddy water were sufficient to kill 100% of dragonfly (*Sympetrum infuscatum*) nymphs in nine days [138].

4.1.4. Insect growth regulators

There is little information about the effect of systemic chitin inhibitors on non-target organisms. Obviously these compounds are harmless to fish at levels above 1 mg/L for a week-long exposures [290], and to all vertebrates in general. IGRs affect mainly the larval stages of Lepidoptera, Coleoptera and Hymenoptera, and their activity last longer than that of other pest control products [178]. The effectiveness of these compounds in controlling target pests is demonstrated by comparing the dietary LC50 of hexaflumuron (0.31 mg/L) to the target cotton worm (*Helicoverpa* sp.), which is 35 times lower than that of the systemic carbamate thiodicarb and less damaging to non-target predators [64]. Aquatic communities of non-target arthropods in rice fields (e.g. Cladocera, Copepoda, Odonata, Notonectidae, Coleoptera and Chironomidae taxa) were not affected by teflubenzuron applied at rates to control mosquitoes (5.6 mg/ha), even though this IGR remained active for several weeks during autumn and winter periods [239].

After application of IGRs to a crop, affected insect pests are prey to many species of spiders, some of which are also susceptible to the toxicity of these products, in particular the ground hunter spiders [211]. Larvae and eggs of pests contaminated with systemic IGR are consumed by a number of predators, including earwigs, which undergo secondary poisoning and stop growing beyond the nymph stage [226]. Chitin inhibitors only show effects on the larvae of predatory insects that had consumed treated-prey, not on the adult insects. As a consequence, predatory populations collapse, as it happened with the ladybeetle *Chilocorus nigritus* that fed on citrus red scales (*Aonidiella aurantii*) in African orchards that had been treated with teflubenzuron [177]. Teflubenzuron sprayed at 16.4 g/ha for locust control in Mali did not affect the non-target arthropods in the herb layer, whereas ground-living Collembola, Thysanura, Coleoptera and Lepidoptera larvae were reduced by about 50% [151]. Moreover, teflubenzuron has multigenerational impacts: experiments with springtails exposed to artificial soil contaminated with this IGR showed that the F2 generation suffered significantly from its effects even when only the F0 generation had been exposed for 10 days [42]. Secondary poisoning with chitin inhibitors can be detrimental also to parasitoids such as *Diadegma semiclausum*, which may fail to produce enough cocoons in the treated hosts, but do not seem to affect the parasitism of other Hymenoptera [98]. For instance, novaluron did not affect the parasitisation of *Trichogramma pretiosum* on mill moth's caterpillars, a pest of tomato crops [44]. On the other hand, teflubenzuron appears to be harmless to predatory mites [32]. IPM programs must always consider the implications of using systemic chitin inhibitors to control specific pests without destroying their natural predators in the first place.

Halofenozide does not appear to cause any acute, adverse effects through topical, residual, or dietary exposure of the ground beetle *Harpalus pennsylvanicus*. In contrast to the negative effects of other systemic insecticides (i.e. imidacloprid), the viability of eggs laid by females fed halofenozide-treated food once, or continuously for 30 days, was not reduced [156].

4.2. Sublethal effects

Very often, sublethal effects of systemic insecticides are a first step towards mortality, as they are caused by the same neurotoxic mechanisms. Apart from these, there may be other effects on reproduction, growth, longevity, etc. when organisms are exposed to low, sublethal doses or concentrations. These effects are only observable in individuals that survive the initial exposure, or in species that are tolerant to insecticides. For a review see [69].

4.2.1. Acetylcholinesterase inhibitors

Longevity of the parasitoid *Microplitis croceipes* that fed on nectar from cotton treated with aldicarb was affected for at least 10 days after application, and its foraging ability of the parasitoid's host was severely impaired for 18 days [257]. Carbofuran caused a significant reduction of adult weight and longevity of the predator ladybug *Hippodamia undecimnotata*, as well as a 55% reduction in fecundity when fed on aphids contaminated with this insecticide [206]. Longevity and survival of *Aphidius ervi*, an important parasitoid of the pea aphid (*Acyrthosiphon pisum*), were significantly reduced after treating with LC25 concentrations of dimethoate or pirimicarb [11]. A significant reduction in body size of females of the predator carabid *Pterostichus melas italicus* and altered sexual dimorphism were observed after long-term exposure in olives groves treated with dimethoate at a rate that caused 10% mortality after three days [104]. Unlike other insecticides, no behavioural effects of dimethoate or tria-zamate on honey bees were recorded [67].

Earthworms (*Lumbricus terrestris*) experienced significant reduction in growth rate and total protein content after soil applications of aldicarb at LC10 or LC25, but only small amounts of residues were detected in the worms [198]. Aldicarb and phorate can also increase infec-tions by *Rhizoctonia* stem canker in potato fields [280].

A typical pattern of sublethal intoxication was revealed when red-winged blackbirds (*Age-laius phoeniceus*) were exposed to increasing doses of dimethoate: 2 mg/kg b.w. doses pro-duced ataraxia, defecation and diarrhoea; neuromuscular dysfunctions and breathing complications appeared at 3 mg/kg, and by 5 mg/kg muscle paralysis and death occurred. The estimated LC50 was 9.9 mg/kg, and all birds died at doses above 28 mg/kg [38]. Al-though sublethal AChE depression by acephate (25% brain) did not affect the attack behav-iour in American kestrels (*Falco sparverius*) [229], nor did alter breeding behaviour in American robins (*Turdus migratorius*) [65], exposure to 256 mg/kg b.w. acephate impaired the migratory orientation of the white-throated sparrow (*Zonotrichia albicollis*) [295]. Similar-ly, low doses of demeton–S-methyl did not affect starlings (*Sturnus vulgaris*) behaviour [279], but doses of 2.5 mg/kg b.w. of dicrotophos administered to female starlings signifi-cantly reduced their parental care and feeding of nestlings [113]. Carbofuran orally adminis-tered to pigeons (*Columba livia*) had profound effects on flight time, with pigeons falling off the pace of the flock when doses were between 0.5 and 1.0 mg/kg b.w. [36].

AChE activities in adductor muscle were depressed in freshwater mussels (*Elliptio complana-ta*) exposed for 96 h at concentrations as low as 0.1 mg/L and 1.3 mg/L of aldicarb and ace-phate respectively, while increasing the water temperature from 21 to 30 °C resulted in mortality [199]. High AChE inhibition (70%) by acephate was not associated with immobili-ty of *Daphnia magna*, but increasing the concentration of acephate further had a strong detri-mental effect on mobility, suggesting that binding sites other than AChE may be involved in acephate toxicity [222].

Exposure of bluegill fish (*Lepomis macrochirus*) to 30 µg/L carbofuran decreased significantly adenylate parameters in gill, liver, muscle and stomach tissues after 10 days, and then re-turned to normal [128]. Also, concentrations of carbofuran at half the LC50 dose for fathead minnow (*Pimephales promelas*) larvae caused reductions in swimming capacity, increased sensitivity to electric shocks, and a reduction in upper lethal temperature [121]. Enzymes of protein and carbohydrate metabolism were altered (some increased, others decreased) in liver and muscle tissues of the freshwater fish, *Clarias batrachus* when exposed to 7.7 mg/L of carbofuran for six days, recovering later to normal levels [26]. Exposure of guppies (*Brachy-danio rerio*) to half the recommended dose for dimethoate (0.025 µl/L) caused morphological changes in hepatocytes within three days, as well as necrosis and other abnormalities [227]. When exposed to a range of monocrotophos concentrations (0.01-1.0 mg/L), male goldfish (*Carassius auratus*) showed higher levels of 17-β-estradiol and vitollogenin and lower levels of testosterone than normal, interfering with gonadotropin synthesis at the pituitary gland [281]. Eggs of the toad *Bufo melanostictus* exposed to acephate hatched normally, but the tad-poles exhibited deformities such as tail distortions and crooked trunk; decreased pigmentation, peeling of the skin, inactivity, delay in emergence of limbs and completion of metamorphosis were also apparent [103].

Insecticide mixtures can enhance not only the acute but also the sublethal effects. For example, disulfoton together with endosulfan caused cytological and biochemical changes in liver of rainbow trout (*Oncorrhynchus mykiss*), independently of their respective modes of action [13]. Mixtures of aldicarb and other insecticides enhanced significantly the establishment of parasitic lungworm nematodes (*Rhandias ranae*) in leopard frogs (*Rana pipiens*) some 21 days after infection [101], as the frog's immune response was suppressed or altered [51]. Similarly, laboratory rats exposed to sublethal mixtures of aldicarb, methomyl and a herbicide (metribuzin) showed learning impairment, immune response and endocrine changes [215].

4.2.2. Insecticides acting on nAChR

Laboratory experiments have shown a number of abnormalities such as less melanin pigmentation, wavy notochord, crooked trunk, fuzzy somites, neurogenesis defects and vasculature defects in zebrafish (*Danio rerio*) embryos exposed to a range of cartap concentrations. The most sensitive organ was the notochord, which displayed defects at concentrations as low as 25 µg/L [308]. It is obvious that essential enzymatic processes are disturbed during embryo development, among which the inhibition of lysyl oxidase is responsible for the notochord undulations observed.

Imidacloprid does not cause high mortality among eggs or adults of the preparasite nematode *Agamermis unka,* but impairs the ability of the nematode to infect nymphs of the host brown planthopper (*Nilaparvata lugens*) [50]. Contrary to this, a synergistic effect of imidacloprid on reproduction of entomopathogenic nematodes against scarab grubs may increase the likelihood of infection by subsequent generations of nematodes, thereby improving their field persistence and biological potential to control grubs. Acetamiprid and thiamethoxam, however, do not show synergist interactions with nematodes [149]. Imidacloprid at 0.1-0.5 mg/kg dry soil disturbs the burrowing ability of *Allolobophora* spp. earthworms [43], and the highest concentration can also induce sperm deformities in the earthworm *Eisenia fetida* [306]. Reduction in body mass (7-39%) and cast production (42-97%) in *Allolobophora* spp. and *Lumbricus terrestris* have also been observed after 7 days exposure to relevant environmental concentrations of imidacloprid [74]. Residues of imidacloprid in maple leaves from treated forests (3–11 mg/kg) did not affect survival of aquatic leaf-shredding insects or litter-dwelling earthworms. However, feeding rates by aquatic insects and earthworms were reduced, leaf decomposition (mass loss) was decreased, measurable weight losses occurred among earthworms, and aquatic and terrestrial microbial decomposition activity was significantly inhibited, thus reducing the natural decomposition processes in aquatic and terrestrial environments [150].

The dispersal ability of the seven-spotted ladybirds (*Coccinella septempunctata*) sprayed with imidacloprid was compromised, and this may have critical consequences for biological control in IPM schemes [21]. A significant reduction of adult weight and longevity of the ladybug *Hippodamia undecimnotata*, as well as 33% reduction in fecundity were observed when this predatory bug fed on aphids contaminated with imidacloprid [206]. Imidacloprid and fipronil had adverse effects on the immune response of the wolf-spider *Pardosa pseudoannulata*, reducing significantly its phenoloxidase activity, the total number of hemocytes and encapsulation rate [282]; the implications of such effects on this natural enemy of rice pests are unknown. When applied in the egg-larval or pupal stages, acetamiprid or imidacloprid reduced the parasitisation capacity of F1 and F2 generation females of *Trichogramma pretiosum* on mill moth's caterpillars (*Anagasta kuehniella*), a pest of tomato crops [44]. Longevity of females of the parasitoid *Microplitis croceipes* that fed on nectar from imidacloprid-treated cotton was affected for at least 10 days after application, while the parasitoid's host foraging ability was severely affected from day 2 onwards [257]. Exposure of western subterranean termites (*Reticulitermes hesperus*) to acetamiprid (1 mg/kg sand) or imidacloprid also impaired locomotion of termites within 1 hour [230].

Bumble bees (*Bombus terrestris*) interrupt their activity for several hours when exposed to imidacloprid sprayed on plants [132], and soil treatment at the highest recommended doses extended the handling times of *B. impatiens* on the complex flowers [194]. Such an impairment affects the bees foraging behaviour and can result in a decreased pollination, lower reproduction and finally in colony mortality due to a lack of food [193]. Although Franklin et al. [96] found that clothianidin residues of 6 µg/kg in canola pollen reduced the production of queens and increased the number of males in *B. impatiens*, their study did not find significant differences with controls due to a high variability in the results. Larval development in wild bees (*Osmia lignaria* and *Megachile rotundata*) was delayed significantly when fed pollen contaminated with either imidacloprid or clothianidin at 30 or 300 µg/kg [1]. Honey bees are more sensitive to neonicotinoids than bumble bees: at 6 µg/kg, imidacloprid clearly induced a decrease in the proportion of active bees [57], and 50-500 µg/L affect significantly their activity, with bees spending more time near the food source [273]. Other authors found that lower activity of honey bees during the hours following oral exposure to 100-500 µg/L imidacloprid in syrup is transitory [186]. In any case, that may explain the delayed homing behaviour of honey bees exposed to 100 µg/L imidacloprid in syrup and their disappearance at higher doses [34, 304]. Honey bees fed on syrup contaminated with acetamiprid increased their sensitivity to antennal stimulation by sucrose solutions at doses of 1 µg/bee and had impaired long-term retention of olfactory learning at 0.1 µg/bee. Contact exposure at 0.1 and 0.5 µg/bee increased locomotor activity and water-induced proboscis extension reflex but had no effect on behaviour [82]. Similar response was obtained with honey bees exposed to thiomethoxam by contact, having impaired long-term retention of olfactory learning at 1 ng/bee [8]. Winter bees surviving chronic treatment with imidacloprid and its metabolite (5-OH-imidacloprid) had reduced learning performances than in summer: the lowest-effect concentration of imidacloprid was lower in summer bees (12 µg/kg) than in winter bees (48 µg/kg), indicating a greater sensitivity of honey bees behaviour in summer bees compared to winter bees [68].

Honey bees infected with the microsporidian *Nosema ceranae* experienced 7 or 5 times higher mortality than normal when fed syrup contaminated with sublethal doses of thiacloprid (5 mg/L) or fipronil (1 µg/L), respectively [293]. *N. ceranae* is a key factor in the CCD in honey bees [127], and the synergistic effect of these systemic insecticides on *Nosema* is probably its underlying cause [213]. Suppression of the immune system is not restricted to bees, as a massive infection of medaka fish by a protozoan ectoparasite (*Trichodina* spp.) when exposed to imidacloprid in rice mesocosms has been documented [236].

Imidacloprid residues in water as low as 0.1 µg/L are sufficient to reduce head and torax length in mayfly nymphs of *Baetis* and *Epeorus*, whether applied as pulses or in continuous exposures for 20 days [6]. At 1 µg/L the insecticides caused feeding inhibition. However, 12-h pulses induced emergence because of stress, whereas constant exposure reduced survivorship progressively. Also, the aquatic worm *Lumbriculus variegatus* experienced immobility during 4 days when exposed to 0.1-10 µg/L imidacloprid [5].

4.2.3. Fipronil

Apart from the extreme acute toxicity of this insecticide to bees, honey bees fed on sucrose syrup contaminated with fipronil (2 µg/kg) reduced significantly their attendance to the feeder [57]. It has also been demonstrated that sublethal concentrations of this insecticide as low as 0.5 ng/bee, whether orally or topically applied, reduce the learning performance of honey bees and impair their olfactory memory but not their locomotor activity [67, 82]. Furthermore, chronic feeding exposure at 1 µg/kg or 0.01 ng/bee reduced learning and orientation, whilst oral treatment of 0.3 ng/bee reduced the number of foraging trips among the exposed workers [66]. In addition to their activity, honey bees fed with sucrose syrup containing 1 µg/L fipronil increased significantly the mortality of bees infected with the endoparasite *Nosema ceranae*, suggesting a synergistic effect between the insecticide and the pathogen [293]. All these sublethal effects reduce the performance of the hive and help explain the decline in honey bee and wild bee pollinators in many countries [205], although fipronil is not alone in causing this demise – neonicotinoids are equally implicated.

Female zebra finches (*Taeniopygia guttata*) fed with single sublethal doses of fipronil (1, 5, and 10 mg/kg b.w.) failed to hatch 6 out of 7 eggs laid. The only chick born was underdeveloped and had fiprole residues in the brain, liver and adipose tissues. By contrast, 12-day-old chicken eggs injected with fipronil (5.5 to 37.5 mg/kg egg weight) hatched normally although the chicks from the highest dose group showed behavioural and developmental abnormalities [145].

Low residues of fipronil in estuary waters (0.63 µg/L) inhibited reproduction of the copepod *Amphiascus tenuiremis* by 73-89%, and this effect seems to be more prevalent on males than on females [45]. Even lower residue levels (0.22 µg/L) halted egg extrusion by 71%, whereas exposure to 0.42 µg/L nearly eliminated reproduction (94% failure) on this species. Based on these results from chronic and sublethal toxicity, a three-generation Leslie matrix model predicted a 62% decline in population size of *A. tenuiremis* at only 0.16 µg/L [47]. Unlike other insecticides, the stress on *Ceriodaphnia dubia* caused by predatory cues of bluegill fish (*Lepomis macrochirus*) was significantly exacerbated when the cladocerans were exposed to 80-160 µg/L of fipronil [223]; however, these concentrations are much higher than the residue levels usually found in waters [99, 163].

While fipronil applied at the recommended rates in rice fields induces biochemical alterations in carp (*Cyprinus carpio*), such metabolic disturbances do not appear to have any effect
on growth nor mortality of this fish after 90 days exposure at <0.65 µg/L [52]. However,
similar residue levels (<1 µg/L) reduced significantly the growth of adult medaka fish
(*Oryzias latipes*) after two weeks of exposure, as well as growth of their offspring in the
first 35 days, even if residues of fipronil by that time were below the analytical detection
limit (0.01 µg/L) [117].

4.2.4. Insect growth regulators

Longevity of predatory bug *Podisus maculiventris* was reduced after preying on Colorado potato beetles that fed on foliage treated with novaluron at 85 g/ha. Females produced fewer
eggs and their hatching was significantly suppressed, while 5th instars that also preyed on
the beetles failed to moult into adults [62]. Novaluron and hexaflumuron significantly decrease (<30%) the total protists population in the guts of termites (*Reticultermes flavipes*), thus
upsetting their digestive homeostasis [165].

4.3. Indirect effects on populations and communities

Indirect effects result from the dynamics of ecosystems. Thus, applications of granular phorate to soil eliminate most soil invertebrates (see 4.1) except for Enchytraeidae worms, which
increase in large numbers and take over the leaf-litter decomposition function carried out by
the eliminated springtails [300].

Resurgence or induction of pests by altering the prey-predator relationships in favour of the
herbivore species is most common. When carbofuran was applied to corn plantations in Nicaragua, the population levels of the noctuid pest *Spodoptera frugiperda* increased because of
lesser foraging activity by predatory ants [212]. Methomyl eliminated the phytoseiid predatory mite *Metaseiulus occidentalis* for 10 days, thus causing an increase in Pacific spider mites
(*Tetranychus pacificus*) and leafhopper (*Eotetranychus willamettei*) populations in the treated
vineyards [130]. Unexpected outbreaks of a formerly innocuous herbivore mite (*Tetranychus
schoenei*) were observed after imidacloprid applications to elms in Central Park, New York.
A three-year investigation on the outbreaks showed that elimination of its predators and the
enhanced fecundity of *T. schoenei* by this insecticide were responsible for that outcome [268].

The widespread use of insecticides usually tips the ecological balance in favour of herbivore
species. For example, dimethoate sprayed on clover fields indirectly reduced the populations of house mice (*Mus musculus*) in the treated areas as the insect food source was deplet-
ed. However, herbivore species such as prairie voles (*Microtus ochrogaster*) and prairie deer
mouse (*Peromyscus maniculatus*) increased in density levels [24], since they had more clover
available due to either higher clover yields or through less competition with the house mice
or both.

A reduction in arthropod populations often implies starvation of insectivorous animals. For
example, densities of two species of lizards and hedgehogs in Madagascar were reduced
45-53% after spraying with fipronil to control a locust outbreak, because their favourite ter-

mite prey was almost eliminated (80-91%) by this chemical [214]. However, this type of indirect impact is difficult to observe and measure in birds, since they can move to other areas or change their resource diet. For example, hemlock forests treated with imidacloprid to control hemlock woolly adelgid (*Adelges tsugae*) reduced significantly Hemiptera and larval Lepidoptera, but not other insect taxa. Although larval Lepidoptera are the primary prey for insectivorous foliage-gleaning birds, many birds were able to find other food resources in the mixed hemlock-deciduous stands that were not treated [87]. Similarly, post-treatment with fipronil for grasshopper control in Wyoming did not affect bird densities, perhaps due to the large initial insect populations; fipronil plots generally had higher avian population densities (nongregarious, insectivores and total birds) than other areas treated with carbaryl [203]. Although some early studies found that fipronil did not have much impact on aquatic communities of Sahelian ponds [158], nor in predatory invertebrates in the Camargue marshes, herons in the latter region avoid rice fields treated with fipronil because of the scarcity of invertebrate food in there [188].

Food aversion to pesticide-treated seeds or plants is a mechanism that may indirectly ameliorate the toxic effects of systemic insecticides such as carbofuran in mice and other small rodents [170]. Some Collembola species (i.e. *Folsomia fimetaria*) avoid dimethoate sprayed areas [86], and female parasitoids (*Cotesia vestalis*) are discouraged from getting to their host –the diamond-back month (*Plutella xylostella*) – in turnip plants treated with methomyl, whereas clothianidin does not produce aversion [248]. Equally, dimethoate and oxydemeton-methyl sprayed on peach trees discourage honey bees from visiting in the first two days after application, while treatments with imidacloprid, acetamiprid and thiamethoxam allow honey bees visits [246]. This helps explain the high long-term impact of neonicotinoids on bees compared to the effect of OP insecticides, even if imidacloprid at high experimental concentrations in syrup (>0.5 mg/L) may also have repellent effect on honey bees [34].

5. Risk assessment of systemic insecticides

All systemic compounds have effects with time of exposure. However, only the persistent chemicals (fipronil, neonicotinoids, cartap and some OPs) have cumulative effects over time, since the non-persistent compounds are quickly degraded in soil and water.

For risk assessment of these compounds it is important to understand their chronic impacts. Unlike traditional protocols based on acute toxicity, the persistent activity of the parent and toxic metabolites requires that exposure time must be taken into consideration [115]. Concerns about the impacts of dietary feeding on honey bees and other non-target organisms are thus justified [9, 60, 228], because the accumulation of small residue levels ingested repeatedly over time will eventually produce a delayed toxic effect [276]. For example, bees that feed on contaminated nectar and pollen from the treated crops are exposed to residues of imidacloprid and fipronil in the range 0.7-10 µg/kg and 0.3-0.4 µg/kg respectively [33], which appear in 11% and 48% of the pollen surveyed in France [48]. Based on those findings an estimate of the predicted environmental concentrations that bees are ingesting in that country can be made for each insecticide. Since there is a log-to-log linear relationship between concentration and time of exposure [234], the critical levels of residue and time of exposure can be determined.

The declining populations of predatory and parasitic arthropods after exposure to recommended applications of most systemic insecticides are worrying. In view of the above, it not so much the small concentrations they are exposed to but the time of exposure that makes the population decline progressively over weeks, months and even years of treatment, as described in this chapter. Lethal and sublethal effects on reproduction are equally implicated. This is the reason why systemic insecticides should be evaluated very carefully before using them in IPM schemes. Obviously, recovery rates are essential for the populations affected to come back, and this usually occurs by recolonisation and immigration of individuals from non-affected areas. For example, modelling based on recovery data after dimethoate application to wheat fields [277] demonstrates that a non-target organism that is reduced by only 20% but is unable to recover is likely to be far more at risk from exposure to a pesticide than an organism that is reduced 99% for a short period but has a higher recovery potential.

The above is also relevant to the impact of small residues of those systemic insecticides that have cumulative effects (e.g. neonicotinoids, fipronil and cartap) on aquatic ecosystems. Because of the short life-cycle of many zooplankton species, the negative population parameters that result from sublethal and chronic effects on such organisms can lead their local populations to extinction [260]. Immediate reductions in populations and species may not always be apparent due to the small residue concentrations and the delayed effects they cause. For example, in recent surveys of pesticide residues in freshwaters of six metropolitan areas of USA, fipronil appears regularly in certain states [254]. Fipronil and its desulfinyl, sulfide, and sulfone degradates were detected at low levels (≤ 0.18–$16\,\mu g/L$) in estuary waters of Southern California [163], and make some 35% of the residues found in urban waters, with a median level of 0.2-$0.44\,\mu g/L$, most frequently during the spring-summer season [99]. Imidacloprid was detected in 89% of water samples in agricultural areas of California, with 19% exceeding the US Environmental Protection Agency's chronic invertebrate Aquatic Life Benchmark of $1.05\,\mu g/L$ [261]. In the Netherlands, imidacloprid appeared in measurable quantities in 30% of the 4,852 water samples collected between 1998 and 2007 [287]. These figures indicate there is already a widespread contamination of waterways and estuaries with persistent systemic insecticides.

The first consequence of such contamination is the progressive reduction, and possible elimination, of entire populations of aquatic arthropods from the affected areas. As time is a critical variable in this type of assessment, it is envisaged that should this contamination continue at the current pace over the years to come the biodiversity and functionality of many aquatic ecosystems will be seriously compromised [191]. Secondly, as these organisms are a primary food source of a large number of vertebrates (e.g. fish, frogs and birds), the depletion of their main food resource will inevitably have indirect impacts on the animal populations that depend on them for their own survival. The case of the partridge in England is an example of how a combination of herbicides and insecticides can bring the demise of a non-target species by indirectly suppressing its food requirements [217]. Therefore, warnings about the possible role of environmental contamination with neonicotinoids in steeply declining populations of birds, frogs, hedgehogs, bats and other insectivorous animals are not far fetched and should be taken seriously [275].

6. Conclusions

This review has brought some light on the direct, sublethal and indirect effects that systemic insecticides have on species populations and ecosystems. Some long-term impacts have been known for some time (e.g. carbofuran, phorate), but it is the rapid increase in the usage of neonicotinoids and other systemic products that poses a new challenge to the ecological risk assessment of agrochemicals. Indeed, current risk protocols, based on acute, short-term toxic effects are inadequate to cope with the chronic exposure and cumulative, delayed impacts of the new compounds. Awareness of the increasing contamination of the environment with active residues of these chemicals should help regulators and managers to implement new approaches for risk assessment of these substances.

Author details

Francisco Sánchez-Bayo[1*], Henk A. Tennekes[2] and Koichi Goka[3]

*Address all correspondence to: sanchezbayo@mac.com

1 University of Technology Sydney, Australia

2 Experimental Toxicology Services (ETS) Nederland BV, The Netherlands

3 National Institute for Environmental Sciences, Japan

References

[1] Abbott, V. A., Nadeau, J. L., Higo, H. A., & Winston, M. L. (2008). Lethal and sublethal effects of imidacloprid on *Osmia lignaria* and clothianidin on *Megachile rotundata* (Hymenoptera: Megachilidae). *J. Econ. Entomol.*, 101(3), 784-796.

[2] Agritox. (2002). Liste des substances actives. *Paris, France: Institut National de la Recherche Agronomique.*

[3] Al-Antary, T. M., Ateyyat, M. A., & Abussamin, B. M. (2010). Toxicity of certain insecticides to the parasitoid *Diaeretiella rapae* (Mcintosh) (Hymenoptera: Aphidiidae)

and its host, the cabbage aphid *Brevicoryne brassicae L.* (Homoptera: Aphididae). *Aust. J. Basic Appl. Sci.*, 4(6), 994-1000.

[4] Al-Haifi, M. A., Khan, M. Z., Murshed, V. A., & Ghole, S. (2006). Effect of dimethoate residues on soil micro-arthropods population in the valley of Zendan, Yemen. *J. Appl. Sci. Environ. Manage.*, 10(2), 37-41.

[5] Alexander, A. C., Culp, J. M., Liber, K., & Cessna, A. J. (2007). Effects of insecticide exposure on feeding inhibition in mayflies and oligochaetes. *Environ. Toxicol. Chem.*, 26(8), 1726-1732.

[6] Alexander, A. C., Heard, K. S., & Culp, J. M. (2008). Emergent body size of mayfly survivors. *Freshwat. Biol.*, 53(1), 171-180.

[7] Ali, A., & Stanley, B. H. (1982). Effects of a new carbamate insecticide, Larvin (UC-51762), on some nontarget aquatic invertebrates. *Florida Entomol.*, 65(4), 477-483.

[8] Aliouane, Y., El -Hassani, A. K., Gary, V., Armengaud, C., Lambin, M., & Gauthier, M. (2009). Subchronic exposure of honeybees to sublethal doses of pesticides: effects on behavior. *Environ. Toxicol. Chem.*, 28(1), 113-122.

[9] Alix, A., & Vergnet, C. (2007). Risk assessment to honey bees: a scheme developed in France for non-sprayed systemic compounds. *Pest Manage. Sci.*, 63(11), 1069-1080.

[10] Anoop, K., & Ram, S. (2012). Effect of biopesticides and insecticides on aphid population, bee visits and yield of mustard. *Ann. Plant Protect. Sci.*, 20(1), 206-209.

[11] Araya, J. E., Araya, M., & Guerrero, M. A. (2010). Effects of some insecticides applied in sublethal concentrations on the survival and longevity of *Aphidius ervi* (Haliday) (Hymenoptera: Aphidiidae) adults. *Chilean J. Agric. Res.*, 70(2), 221-227.

[12] Armbrust, K. L., & Peeler, H. B. (2002). Effects of formulation on the run-off of imidacloprid from turf. *Pest Manage. Sci.*, 58(7), 702-706.

[13] Arnold, H., Pluta, H. J., & Braunbeck, T. (1995). Simultaneous exposure of fish to endosulfan and disulfoton in vivo: ultrastructural, stereological and biochemical reactions in hepatocytes of male rainbow trout (*Oncorhynchus mykiss*). *Aquat. Toxicol.*, 33, 17-34.

[14] Ateyyat, M. (2012). Selectivity of four insecticides to woolly apple aphid, Eriosoma lanigerum (Hausmann) and its sole parasitoid, *Aphelinus mali* (Hald.). *World Appl. Sci. J.*, 16(8), 1060-1064.

[15] Azizullah, A., Richter, P., & Häder, D. P. (2011). Comparative toxicity of the pesticides carbofuran and malathion to the freshwater flagellate *Euglena gracilis*. *Ecotoxicology*, 20(6), 1442-1454.

[16] Babu, B. S., & Gupta, G. P. (1986). Effect of systemic insecticides on the population of soil arthropods in a cotton field. *J. Soil Biol. Ecol.*, 6(1), 32-41.

[17] Babu, B. S., & Gupta, G. P. (1988). Efficacy of systemic insecticides against cotton jassid (*Amrasca devastans*) and their effect on non-target organisms in upland cotton (*Gossypium hirsutum*). *Indian J. Agric. Sci.*, 58(6), 496-499.

[18] Balanca, G., & Visscher, M. N.d. (1997). Effects of very low doses of fipronil on grasshoppers and non-target insects following field trials for grasshopper control. *Crop Protection*, 16, 553-564.

[19] Balcomb, R., Bowen, C. I., Wright, D., & Law, M. (1984). Effects on wildlife of at-planting corn applications of granular carbofuran. *J. Wildl. Manage.*, 48(4), 1353-1359.

[20] Balconi, C., Mazzinelli, G., & Motto, M. (2011). Neonicotinoid insecticide seed coatings for the protection of corn kernels and seedlings, and for plant yield. *Maize Genetics Cooperation Newsletter* [84], 3.

[21] Banks, J. E., & Stark, J. D. (2011). Effects of a nicotinic insecticide, imidacloprid, and vegetation diversity on movement of a common predator, *Coccinella septempunctata* *Biopestic. Int.* 7(2), 113-122.

[22] Barahona, M. V., & Sánchez-Fortún, S. (1999). Toxicity of carbamates to the brine shrimp *Artemia salina* and the effect of atropine, BW284c51, iso-OMPA and 2-PAM on carbaryl toxicity. *Environ. Pollut.*, 104(3), 469-476.

[23] Barbee, G. C., & Stout, M. J. (2009). Comparative acute toxicity of neonicotinoid and pyrethroid insecticides to non-target crayfish (*Procambarus clarkii*) associated with rice-crayfish crop rotations. *Pest Manage. Sci.*, 65(11), 1250-1256.

[24] Barrett, G. W., & Darnell, R. M. (1967). Effects of dimethoate on small mammal populations. *Am. Midland Nat.*, 77, 164-175.

[25] Baur, M. E., Ellis, J., Hutchinson, K., & Boethel, D. J. (2003). Contact toxicity of selective insecticides for non-target predaceous hemipterans in soybeans. *J. Entomol. Sci.*, 38(2), 269-277.

[26] Begum, G. (2004). Carbofuran insecticide induced biochemical alterations in liver and muscle tissues of the fish *Clarias batrachus* (linn) and recovery response. *Aquat. Toxicol.*, 66(1), 83-92.

[27] Beketov, M., Schafer, R. B., Marwitz, A., Paschke, A., & Liess, M. (2008). Long-term stream invertebrate community alterations induced by the insecticide thiacloprid: effect concentrations and recovery dynamics. *Sci. Total Environ.*, 405, 96-108.

[28] Beketov, M. A., & Liess, M. (2008). Acute and delayed effects of the neonicotinoid insecticide thiacloprid on seven freshwater arthropods. *Environ. Toxicol. Chem.*, 27(2), 461-470.

[29] Belfroid, A. C., van Drunen, M., Beek, M. A., Schrap, S. M., van Gestel, C. A. M., & van Hattum, B. (1998). Relative risks of transformation products of pesticides for aquatic ecosystems. *Sci. Total Environ.*, 222(3), 167-183.

[30] Bellows, T. S., Morse, J. G., Gaston, L. K., & Bailey, J. B. (1988). The fate of two systemic insecticides and their impact on two phytophagous and a beneficial arthropod in a citrus agroecosystem. *J. Econ. Entomol.*, 81(3), 899-904.

[31] Berg, H. (2001). Pesticide use in rice and rice-fish farms in the Mekong Delta, Vietnam. *Crop Protection*, 20(10), 897-905.

[32] Bluemel, S., & Stolz, M. (1993). Investigations on the effect of insect growth regulators and inhibitors on the predatory mite *Phytoseiulus persimilis* A.H. with particular emphasis on cyromazine. *Zeitschrift fuer Pflanzenkrankheiten und Pflanzenschutz*, 100(2), 150-154.

[33] Bonmatin, J. M., Marchand, P. A., Cotte, J. F., Aajoud, A., Casabianca, H., Goutailler, G., & Courtiade, M. (2007). Bees and systemic insecticides (imidacloprid, fipronil) in pollen: subnano-quantification by HPLC/MS/MS and GC/MS. *Environmental fate and ecological effects of pesticides, Re, A.A.M.d. et al., editors*, 827-834, 978-8-87830-473-4.

[34] Bortolotti, L., Montanari, R., Marcelino, J., Medrzycki, P., Maini, S., & Porrini, C. (2003). Effects of sublethal imidacloprid doses on the homing rate and foraging activity of honey bees. *Bull. Insectology*, 56(1), 63-67.

[35] Bottger, R., Schaller, J., & Mohr, S. (2012). Closer to reality - the influence of toxicity test modifications on the sensitivity of *Gammarus roeseli* to the insecticide imidacloprid. *Ecotoxicol. Environ. Saf.*, 81(0), 49-54.

[36] Brasel, J., Collier, A., & Pritsos, C. (2007). Differential toxic effects of carbofuran and diazinon on time of flight in pigeons (*Columba livia*): potential for pesticide effects on migration. *Toxicol. Appl. Pharmacol.*, 219(2-3), 241-246.

[37] Bringolf, R. B., Cope, W. G., Eads, C. B., Lazaro, P. R., Barnhart, M. C., & Shea, D. (2007). Acute and chronic toxicity of technical-grade pesticides to glochidia and juveniles of freshwater mussels (Unionidae). *Environ. Toxicol. Chem.*, 26(10), 2094-2100.

[38] Brunet, R., Girard, C., & Cyr, A. (1997). Comparative study of the signs of intoxication and changes in activity level of red-winged blackbirds (*Agelaius phoeniceus*) exposed to dimethoate. *Agric. Ecosyst. Environ.*, 64, 201-209.

[39] Brunetto, R., Burguera, M., & Burguera, J. L. (1992). Organophosphorus pesticide residues in some watercourses from Merida, Venezuela. *Sci. Total Environ.*, 114, 195-204.

[40] Buckingham, S., Lapied, B., Corronc, H., & Sattelle, F. (1997). Imidacloprid actions on insect neuronal acetylcholine receptors. *J. Exp. Biol.*, 200(21), 2685-2692.

[41] Bunyan, P. J., Heuvel, M. J.v.d, Stanley, P. I., & Wright, E. N. (1981). An intensive field trial and a multi-site surveillance exercise on the use of aldicarb to investigate methods for the assessment of possible environmental hazards presented by new pesticides. *Agro-Ecosystems*, 7(3), 239-262.

[42] Campiche, S., L'Arnbert, G., Tarradellas, J., & Becker-van, Slooten. K. (2007). Multi-generation effects of insect growth regulators on the springtail *Folsomia candida*. *Eco-toxicol. Environ. Saf.*, 67(2), 180-189.

[43] Capowiez, Y., Bastardie, F., & Costagliola, G. (2006). Sublethal effects of imidacloprid on the burrowing behaviour of two earthworm species: modifications of the 3D bur-row systems in artificial cores and consequences on gas diffusion in soil. *Soil Biol. Bio-chem.*, 38(2), 285-293.

[44] Carvalho, G. A., Godoy, M. S., Parreira, D. S., & Rezende, D. T. (2010). Effect of chem-ical insecticides used in tomato crops on immature *Trichogramma pretiosum* (Hyme-noptera: Trichogrammatidae). *Revista Colombiana de Entomologia*, 36(1), 10-15.

[45] Cary, T. L., Chandler, G. T., Volz, D. C., Walse, S. S., & Ferry, J. L. (2003). Phenylpyra-zole insecticide fipronil induces male infertility in the estuarine meiobenthic crusta-cean *Amphiascus tenuiremis*. *Environ. Sci. Technol.*, 38(2), 522-528.

[46] Chai, L. K., Wong, M. H., Mohd-Tahir, N., & Hansen, H. C. B. (2010). Degradation and mineralization kinetics of acephate in humid tropic soils of Malaysia. *Chemo-sphere*, 79(4), 434-440.

[47] Chandler, G. T., Cary, T. L., Volz, D. C., Walse, S. S., Ferry, J. L., & Klosterha, S. L. (2004). Fipronil effects on estuarine copepod (*Amphiascus tenuiremis*) development, fertility, and reproduction: a rapid life-cycle assay in 96-well microplate format. *Envi-ron. Toxicol. Chem.*, 23(1), 117-124.

[48] Chauzat, M., , P., Martel, A. C., Cougoule, N., Porta, P., Lachaize, J., Zeggane, S., Au-bert, M., Carpentier, P., Faucon, J., & , P. (2011). An assessment of honeybee colony matrices, *Apis mellifera* (Hymenoptera: Apidae) to monitor pesticide presence in con-tinental France. *Environ. Toxicol. Chem.*, 30(1), 103-111.

[49] Chen, X. D., Culbert, E., Hebert, V., & Stark, J. D. (2010). Mixture effects of the nonyl-phenyl polyethoxylate, R-11 and the insecticide, imidacloprid on population growth rate and other parameters of the crustacean, *Ceriodaphnia dubia*. *Ecotoxicol. Environ. Saf.*, 73(2), 132-137.

[50] Choo, H. Y., Kim, H. H., & Kaya, H. K. (1998). Effects of selected chemical pesticides on Agamermis unka (Nematoda: Mermithidae), a parasite of the brown plant hop-per, Nilaparvata lugens. Biocontrol Sci. Technol. , 8(3), 413-427.

[51] Christin, M. S., Gendron, A. D., Brousseau, P., Ménard, L., Marcogliese, D. J., Cyr, D., Ruby, S., & Fournier, M. (2003). Effects of agricultural pesticides on the immune sys-tem of *Rana pipiens* and on its resistance to parasitic infection. *Environ. Toxicol. Chem.*, 22(5), 1127-1133.

[52] Clasen, B., Loro, V. L., Cattaneo, R., Moraes, B., Lopes, T., Avila, L., A.d, , Zanella, R., Reimche, G. B., & Baldisserotto, B. (2012). Effects of the commercial formulation con-taining fipronil on the non-target organism *Cyprinus carpio*: implications for rice-fish cultivation. *Ecotoxicol. Environ. Saf.*, 77, 45-51.

[53] Clements, R. O., Bentley, B. R., & Jackson, C. A. (1986). The impact of granular for-
      mulations of phorate, terbufos, carbofuran, carbosulfan and thiofanox on newly
      sown Italian ryegrass, *Lolium multiflorum. Crop Protection*, 5(6), 389-394.

[54] Cloyd, R. A., & Bethke, J. A. (2011). Impact of neonicotinoid insecticides on natural
      enemies in greenhouse and interiorscape environments. *Pest Manage. Sci.*, 67(1), 3-9.

[55] Cockfield, S. D., & Potter, D. A. (1983). Short-term effects of insecticidal applications
      on predaceous arthropods and oribatid mites in Kentucky blue grass turf. *Environ.
      Entomol.*, 12(4), 1260-1264.

[56] Cole, L. M., Nicholson, R. A., & Casida, J. E. (1993). Action of phenylpyrazole insecti-
      cides at the GABA-gated chloride channel. *Pestic. Biochem. Physiol.*, 46(1), 47-54.

[57] Colin, M. E., Bonmatin, J. M., Moineau, I., Gaimon, C., Brun, S., & Vermandere, J. P.
      (2004). A method to quantify and analyze the foraging activity of honey bees: rele-
      vance to the sublethal effects induced by systemic insecticides. *Arch. Environ. Contam.
      Toxicol.*, 47(3), 387-395.

[58] Colinas, C., Ingham, E., & Molina, R. (1994). Population responses of target and non-
      target forest soil organisms to selected biocides. *Soil Biol. Biochem.*, 26(1), 41-47.

[59] Cortet, J., & Poinsot-Balaguer, N. (2000). Impact of phytopharmaceutical products on
      soil microarthropods in an irrigated maize field: The use of the litter bag method.
      Can. J. Soil Sci., 80(2), 237-249.

[60] Cresswell, J. (2011). A meta-analysis of experiments testing the effects of a neonicoti-
      noid insecticide (imidacloprid) on honey bees. *Ecotoxicology*, 20(1), 149-157.

[61] Cutler, G. C., & Scott-Dupree, C. D. (2007). Exposure to clothianidin seed-treated can-
      ola has no long-term impact on honey bees. *J. Econ. Entomol.*, 100(3), 765-772.

[62] Cutler, G. C., Scott-Dupree, C. D., Tolman, J. H., & Harris, C. R. (2006). Toxicity of the
      insect growth regulator novaluron to the non-target predatory bug *Podisus maculi-
      ventris* (Heteroptera: Pentatomidae). *Biol. Control*, 38(2), 196-204.

[63] Dai, Y., Ji, W., Chen, T., Zhang, W., Liu, Z., Ge, F., & Yuan, S. (2010). Metabolism of
      the neonicotinoid insecticides acetamiprid and thiacloprid by the yeast *Rhodotorula
      mucilaginosa* strain IM-2. *J. Agric. Food Chem.*, 58(4), 2419-2425.

[64] Dastjerdi, H. R., Hejazi, M. J., Ganbalani, G. N., & Saber, M. (2008). Toxicity of some
      biorational and conventional insecticides to cotton bollworm, *Helicoverpa armigera*
      (Lepidoptera: Noctuidae) and its ectoparasitoid, *Habrobracon hebetor* (Hymenoptera:
      Braconidae). *J. Entomol. Soc. Iran*, 28(1), 27-37.

[65] Decarie, R., Des, Granges. J. L., Lepine, C., & Morneau, F. (1993). Impact of insecti-
      cides on the American robin (*Turdus migratorius*) in a suburban environment. *Envi-
      ron. Pollut.*, 80, 231-238.

[66] Decourtye, A., Devillers, J., Aupinel, P., Brun, F., Bagnis, C., Fourrier, J., & Gauthier, M. (2011). Honeybee tracking with microchips: a new methodology to measure the effects of pesticides. *Ecotoxicology*, 20(2), 429-437.

[67] Decourtye, A., Devillers, J., Genecque, E., Menach, K. L., Budzinski, H., Cluzeau, S., & Pham-Delègue, M. H. (2005). Comparative sublethal toxicity of nine pesticides on olfactory learning performances of the honeybee *Apis mellifera*. *Arch. Environ. Contam. Toxicol.*, 48(2), 242-250.

[68] Decourtye, A., Lacassie, E., & Pham-Delègue, M. H. (2003). Learning performances of honeybees (*Apis mellifera* L) are differentially affected by imidacloprid according to the season. *Pest Manage. Sci.*, 59(3), 269-278.

[69] Desneux, N., Decourtye, A., Delpuech, J., & , M. (2007). The sublethal effects of pesticides on beneficial arthropods. *Annu. Rev. Entomol.*, 52, 81-106.

[70] Dewar, A. M., Haylock, L. A., Garner, B. H., Sands, R. J. N., & Pilbrow, J. (2005). Neonicotinoid seed treatments - the panacea for most pest problems in sugar beet. *Aspects Appl. Biol.* [76], 3-12.

[71] Dhadialla, T. S., Carlson, G. R., & Le , D. P. (1998). New insecticides with ecdysteroidal and juvenile hormone activity. *Annu. Rev. Entomol.*, 43(1), 545-569.

[72] Dieter, C. D., Duffy, W. G., & Flake, L. D. (1996). The effect of phorate on wetland macroinvertebrates. *Environ. Toxicol. Chem.*, 15(3), 308-312.

[73] Dieter, C. D., Flake, L. D., & Duffy, W. G. (1995). Effects of phorate on ducklings in northern prairie wetlands. *J. Wildl. Manage.*, 59(3), 498-505.

[74] Dittbrenner, N., Triebskorn, R., Moser, I., & Capowiez, Y. (2010). Physiological and behavioural effects of imidacloprid on two ecologically relevant earthworm species (*Lumbricus terrestris* and *Aporrectodea caliginosa*). *Ecotoxicology*, 19, 1567-1573.

[75] Drescher, W., & Geusen-Pfister, H. (1991). Comparative testing of the oral toxicity of acephate, dimethoate and methomyl to honeybees, bumblebees and Syrphidae. *Acta Horticulturae*, 288, 133-138.

[76] Easterbrook, M. A. (1997). A field assessment of the effects of insecticides on the beneficial fauna of strawberry. *Crop Protection*, 16(2), 147-152.

[77] Ecoli, C. C., Moraes, J. C., & Vilela, M. (2010). Suplementos alimentares e isca toxica no manejo do bicho-mineiro e de seus inimigos naturais. *Coffee Sci.*, 5(2), 167-172.

[78] Eisenback, B. M., Salom, S. M., Kok, L. T., & Lagalante, A. F. (2010). Lethal and sublethal effects of imidacloprid on hemlock woolly adelgid (Hemiptera: Adelgidae) and two introduced predator species. *J. Econ. Entomol.*, 103(4), 1222-34.

[79] Eisenhauer, N., Klier, M., Partsch, S., Sabais, A. C. W., Scherber, C., Weisser, W. W., & Scheu, S. (2009). No interactive effects of pesticides and plant diversity on soil microbial biomass and respiration. *Appl. Soil Ecol.*, 42(1), 31-36.

[80] El -Din, H. A. S., & Girgis, N. R. (1997). Susceptibility of honey bee workers, *Apis mellifera* L. to nine different insecticides. *Ann. Agric. Sci. Moshtohor*, 35(4), 2571-2582.

[81] El -Hassani, A. K., Dacher, M., Gary, V., Lambin, M., Gauthier, M., & Armengaud, C. (2008). Effects of sublethal doses of acetamiprid and thiamethoxam on the behavior of the honeybee (*Apis mellifera*). *Arch. Environ. Contam. Toxicol.*, 54(4), 653-661.

[82] El -Hassani, A. K., Dacher, M., Gauthier, M., & Armengaud, C. (2005). Effects of sublethal doses of fipronil on the behavior of the honeybee (*Apis mellifera*). *Pharmacol. Biochem. Behavior*, 82(1), 30-39.

[83] Elbert, A., Haas, M., Springer, B., Thielert, W., & Nauen, R. (2008). Applied aspects of neonicotinoid uses in crop protection. *Pest Manage. Sci.*, 64(11), 1099-1105.

[84] Elliott, J. E., Wilson, L. K., Langelier, K. M., Mineau, P., & Sinclair, P. H. (1997). Secondary poisoning of birds of prey by the organophosphorus insecticide, phorate. *Ecotoxicology*, 6(4), 219-231.

[85] Endlweber, K., Schädler, M., & Scheu, S. (2005). Effects of foliar and soil insecticide applications on the collembolan community of an early set-aside arable field. *Appl. Soil Ecol.*, 31(1-2), 136-146.

[86] Fabian, M., & Petersen, H. (1994). Short-term effects of the insecticide dimethoate on activity and spatial distribution of a soil inhabiting collembolan *Folsomia fimetaria* Linne (Collembola:Isotomidae). *Pedobiologia*, 38(4), 289-302.

[87] Falcone, J. F., & De Wald, L. E. (2010). Comparisons of arthropod and avian assemblages in insecticide-treated and untreated eastern hemlock (*Tsuga canadensis* L. Carr) stands in Great Smoky Mountains National Park, USA. *Forest Ecol. Manage.*, 260(5), 856-863.

[88] Farinos, G. P., de la Poza, M., Hernandez-Crespo, P., Ortego, F., & Castanera, P. (2008). Diversity and seasonal phenology of aboveground arthropods in conventional and transgenic maize crops in Central Spain. *Biol. Control*, 44(3), 362-371.

[89] Faucon, J. P., Aurières, C., Drajnudel, P., Mathieu, L., Ribière, M., Martel, A. C., Zeggane, S., Chauzat, M. P., & Aubert, M. F. A. (2005). Experimental study on the toxicity of imidacloprid given in syrup to honey bee (*Apis mellifera*) colonies. *Pest Manage. Sci.*, 61(2), 111-125.

[90] Fernandes, M., E.d, S., Fernandes, F. L., Picanco, M. C., Queiroz, R. B., Silva, R. S.d, & Huertas, A. A. G. (2008). Physiological selectivity of insecticides to *Apis mellifera* (Hymenoptera: Apidae) and *Protonectarina sylveirae* (Hymenoptera: Vespidae) in citrus. *Sociobiology*, 51(3), 765-774.

[91] Fleming, W., & Bradbury, S. (1981). Recovery of cholinesterase activity in mallard ducklings administered organophosphorus pesticides. *J. Toxicol. Environ. Health B*, 8(5-6), 885-97.

[92] Flickinger, E. L., White, D. H., Mitchell, C. A., & Lamont, T. G. (1984). Monocrotophos and dicrotophos residues in birds as a result of misuse of organophosphates in Matagorda County, Texas. *J.A.O.A.C.*, 67, 827-828.

[93] Fluetsch, K. M., & Sparling, D. W. (1994). Avian nesting success and diversity in conventionally and organically managed apple orchards. *Environ. Toxicol. Chem.*, 13(10), 1651-1659.

[94] Fowle, C. D. (1966). The effects of phosphamidon on birds in New Brunswick forests. *J. Appl. Ecol.*, 3, 169-170.

[95] Frampton, G. K., Brink, P., & J.v.d, . (2007). Collembola and macroarthropod community responses to carbamate, organophosphate and synthetic pyrethroid insecticides: direct and indirect effects. *Environ. Pollut.*, 147(1), 14-25.

[96] Franklin, M. T., Winston, M. L., & Morandin, L. A. (2004). Effects of clothianidin on *Bombus impatiens* (Hymenoptera: Apidae) colony health and foraging ability. *J. Econ. Entomol.*, 97(2), 369-373.

[97] Freuler, J., Blandenier, G., Meyer, H., & Pignon, P. (2001). Epigeal fauna in a vegetable agroecosystem. *Mitteilungen der Schweizerischen Entomologischen Gesellschaft*, 74(1-2), 17-42.

[98] Furlong, M. J., Verkerk, R. H. J., & Wright, D. J. (1994). Differential effects of the acylurea insect growth regulator teflubenzuron on the adults of two endolarval parasitoids of *Plutella xylostella, Cotesia plutellae* and *Diadegma semiclausum*. *Pestic. Sci.* , 41(4), 359-364.

[99] Gan, J., Bondarenko, S., Oki, L., Haver, D., & Li, J. X. (2012). Occurrence of fipronil and its biologically active derivatives in urban residential runoff. *Environ. Sci. Technol.*, 46(3), 1489-1495.

[100] Gao, J., Garrison, A. W., Hoehamer, C., Mazur, C. S., & Wolfe, N. L. (2000). Uptake and phytotransformation of organophosphorus pesticides by axenically cultivated aquatic plants. *J. Agric. Food Chem.*, 48, 6114-6120.

[101] Gendron, A. D., Marcogliese, D. J., Barbeau, S., Christin, M. S., Brousseau, P., Ruby, S., Cyr, D., & Fournier, M. (2003). Exposure of leopard frogs to a pesticide mixture affects life history characteristics of the lungworm *Rhabdias ranae*. *Oecologia*, 135, 469-476.

[102] Georgiadis, P. T., Pistorius, J., & Heimbach, U. (2010). Vom Winde verweht- Abdrift von Beizstauben- ein Risiko fur Honigbienen (*Apis mellifera* L.)? *Julius-Kuhn-Archiv* [424], 33.

[103] Ghodageri, M. G., & Katti, P. (2011). Morphological and behavioral alterations induced by endocrine disrupters in amphibian tadpoles. *Toxicol. Environ. Chem.*, 93(10), 2012-2021.

[104] Giglio, A., Giulianini, P. G., Zetto, T., & Talarico, F. (2011). Effects of the pesticide dimethoate on a non-target generalist carabid, *Pterostichus melas italicus* (Dejean, 1828) (Coleoptera: Carabidae). *Italian J. Zool.*, 78(4), 471-477.

[105] Girolami, V., Marzaro, M., Vivan, L., Mazzon, L., Greatti, M., Giorio, C., Marton, D., & Tapparo, A. (2012). Fatal powdering of bees in flight with particulates of neonicotinoids seed coating and humidity implication. *J. Appl. Entomol.*, 136(1/2), 17-26.

[106] Goldman, L. R., Smith, D. F., & Neutra, R. R. (1990). Pesticide food poisoning from contaminated watermelons in California, 1985. *Arch. Environ. Health*, 45(4), 229-236.

[107] Golombieski, J. I., Marchesan, E., Baumart, J. S., Reimche, G. B., Junior, C. R., Storck, L., & Santos, S. (2008). Cladocera, Copepoda e Rotifera em rizipiscicultura tratada com os herbicidas metsulfuron-metílico e azimsulfuron e o inseticida carbofuran. *Ciencia Rural*, 38(8), 2097-2102.

[108] Gomez-Eyles, J. L., Svendsen, C., Lister, L., Martin, H., Hodson, M. E., & Spurgeon, D. J. (2009). Measuring and modelling mixture toxicity of imidacloprid and thiacloprid on *Caenorhabditis elegans* and *Eisenia fetida*. *Ecotoxicol. Environ. Saf.*, 72(1), 71-79.

[109] Gour, I. S., & Pareek, B. L. (2005). Relative toxicity of some insecticides to coccinellid, *Coccinella septempunctata* Linn. and Indian honey bee, *Apis cerana indica*. *Indian J. Agric. Res.*, 39(4), 299-302.

[110] Grafton-Cardwell, E. E., & Gu, P. (2003). Conserving vedalia beetle, *Rodolia cardinalis* (Mulsant) (Coleoptera: Coccinellidae), in citrus: A continuing challenge as new insecticides gain registration. *J. Econ. Entomol.*, 96(5), 1388-1398.

[111] Gregorc, A., & Bozic, J. (2004). Is honey bee colonies mortality related to insecticide use in agriculture? [Ali cebelje druzine odmirajo zaradi uporabe insekticida v kmetijstvu?]. *Sodobno Kmetijstvo*, 37(7), 29-32.

[112] Grout, T. G., Richards, G. I., & Stephen, P. R. (1997). Further non-target effects of citrus pesticides on *Euseius addoensis* and *Euseius citri* (Acari: Phytoseiidae). *Exp. Appl. Acarol.*, 21(3), 171-177.

[113] Grue, C. E., Powell, G. V. N., & Mc Chesney, M. J. (1982). Care of nestlings by wild female starlings exposed to an organophosphate pesticide. *J. Appl. Ecol.*, 19, 327-335.

[114] Grue, C. E., & Shipley, B. K. (1984). Sensitivity of nestling and adult starlings to dicrotophos, an organophosphate pesticide. *Environ. Res.*, 35(2), 454-465.

[115] Halm, M. P., Rortais, A., Arnold, G., Taséi, J. N., & Rault, S. (2006). New risk assessment approach for systemic insecticides: the case of honey bees and imidacloprid (Gaucho). *Environ. Sci. Technol.*, 40(7), 2448-2454.

[116] Hawthorne, D. J., & Dively, G. P. (2011). Killing them with kindness? In-hive medications may inhibit xenobiotic efflux transporters and endanger honey bees. *PLoS One*, (November), e26796.

[117] Hayasaka, D., Korenaga, T., Sánchez-Bayo, F., & Goka, K. (2012a). Differences in eco-logical impacts of systemic insecticides with different physicochemical properties on biocenosis of experimental paddy fields. *Ecotoxicology*, 21(1), 191-201.

[118] Hayasaka, D., Korenaga, T., Suzuki, K., Saito, F., Sánchez-Bayo, F., & Goka, K. (2012b). Cumulative ecological impacts of two successive annual treatments of imi-dacloprid and fipronil on aquatic communities of paddy mesocosms. *Ecotoxicol. Environ. Saf.*, 80, 355-362.

[119] Hayasaka, D., Korenaga, T., Suzuki, K., Sánchez-Bayo, F., & Goka, K. (2012c). Differ-ences in susceptibility of five cladoceran species to two systemic insecticides, imida-cloprid and fipronil. *Ecotoxicology*, 21(2), 421-427.

[120] He, Y., Zhao, J., Zheng, Y., Desneux, N., & Wu, K. (2012). Lethal effect of imidaclo-prid on the coccinellid predator *Serangium japonicum* and sublethal effects on preda-tor voracity and on functional response to the whitefly *Bemisia tabaci*. *Ecotoxicology*, 1-10.

[121] Heath, A. G., Joseph, J., Cech, J., Brink, L., Moberg, P., & Zinkl, J. G. (1997). Physio-logical responses of fathead minnow larvae to rice pesticides. *Ecotoxicol. Environ. Saf.*, 37(3), 280-288.

[122] Heinrichs, E. A., Aquino, G. B., Chelliah, S., Valencia, S. L., & Reissig, W. H. (1982). Resurgence of *Nilaparvata lugens* (Stal) populations as influenced by method and tim-ing of insecticide applications in lowland rice. *Environ. Entomol.*, 11(1), 78-84.

[123] Hela, D. G., Lambropoulou, D. A., Konstantinou, I. K., & Albanis, T. A. (2005). Envi-ronmental monitoring and ecological risk assessment for pesticide contamination and effects in Lake Pamvotis, northwestern Greece. *Environ. Toxicol. Chem.*, 24(6), 1548-1556.

[124] Held, D. W., & Parker, S. (2011). Efficacy of soil applied neonicotinoid insecticides against the azalea lace bug, *Stephanitis pyrioides*, in the landscape. *Florida Entomol.*, 94(3), 599-607.

[125] Henry, M.l., Beguin, M., Requier, F., Rollin, O., Odoux, J. F.o, Aupinel, P., Aptel, J., Tchamitchian, S., & Decourtye, A. (2012). A common pesticide decreases foraging success and survival in honey bees. *Science*, 336, 348-350.

[126] Heong, K. L., Escalada, M. M., & Mai, V. (1994). An analysis of insecticide use in rice: case studies in the Philippines and Vietnam. *Int. J. Pest Manage.*, 40(2), 173-178.

[127] Higes, M., Martín-Hernández, R., Martínez-Salvador, A., Garrido-Bailón, E., Gonzá-lez-Porto, A. V., Meana, A., Bernal, J. L., Del Nozal, M. J., & Bernal, J. (2010). A pre-liminary study of the epidemiological factors related to honey bee colony loss in Spain. *Environ. Microbiol. Rep.*, 2(2), 243-250.

[128] Hohreiter, D. W., Reinert, R. E., & Bush, P. B. (1991). Effects of the insecticides carbo-furan and fenvalerate on adenylate parameters in bluegill sunfish (*Lepomis macrochi-rus*). *Arch. Environ. Contam. Toxicol.*, 21(3), 325-331.

[129] Hoshino, T., & Takase, I. (1993). New insecticide imidacloprid - Safety assessment. *Noyaku Kenkyu*, 39(3), 37-45.

[130] Hoy, M. A., Flaherty, D., Peacock, W., & Culver, D. (1979). Vineyard and laboratory evaluations of methomyl, dimethoate and permethrin for a grape pest management program in the San Joaquin Valley of California, USA. *J. Econ. Entomol.*, 72(2), 250-255.

[131] Huusela-Veistola, E. (2000). Effects of pesticide use on non-target arthropods in a Finnish cereal field. *Aspects Appl. Biol.* [62], 67-72.

[132] Incerti, F., Bortolotti, L., Porrini, C., Sbrenna, A. M., & Sbrenna, G. (2003). An extended laboratory test to evaluate the effects of pesticides on bumblebees. Preliminary results. *Bull. Insectology*, 56(1), 159-164.

[133] Ingham, E. R., Coleman, D. C., & Crossley, D. A. Jr. (1994). Use of sulfamethoxazole-penicillin, oxytetracycline, carbofuran, carbaryl, naphthalene and Temik to remove key organism groups in soil in a corn agroecosystem. *J. Sustain. Agric.*, 4(3), 7-30.

[134] Iwasa, T., Motoyama, N., Ambrose, J. T., & Roe, R. M. (2004). Mechanism for the differential toxicity of neonicotinoid insecticides in the honey bee, *Apis mellifera*. *Crop Protection*, 23(5), 371-378.

[135] Jansen, J. P. (2000). A three-year field study on the short-term effects of insecticides used to control cereal aphids on plant-dwelling aphid predators in winter wheat. *Pest Manage. Sci.*, 56(6), 533-539.

[136] Jemec, A., Tisler, T., Drobne, D., Sepcifá, K., Fournier, D., & Trebse, P. (2007). Comparative toxicity of imidacloprid, of its commercial liquid formulation and of diazinon to a non-target arthropod, the microcrustacean *Daphnia magna*. *Chemosphere*, 68(8), 1408-18.

[137] Jeschke, P., Nauen, R., Schindler, M., & Elbert, A. (2010). Overview of the status and global strategy for neonicotinoids. *J. Agric. Food Chem.*, 59(7), 2897-2908.

[138] Jinguji, H., Thuyet, D., Ueda, T., & Watanabe, H. (2012). Effect of imidacloprid and fipronil pesticide application on *Sympetrum infuscatum* (Libellulidae: Odonata) larvae and adults. *Paddy Water Environ.*, online first:, 1-8.

[139] Johansson, M., Piha, H., Kylin, H., & Merilä, J. (2006). Toxicity of six pesticides to common frog (*Rana temporaria*) tadpoles. *Environ. Toxicol. Chem.*, 25(12), 3164-3170.

[140] Johnson, B. T. (1986). Potential impact of selected agricultural chemical contaminants on a northern prairie wetland: a microcosm evaluation. *Environ. Toxicol. Chem.*, 5(5), 473-485.

[141] Joy, V. C., & Chakravorty, P. P. (1991). Impact of insecticides on nontarget microarthropod fauna in agricultural soil. *Ecotoxicol. Environ. Saf.*, 22(1), 8-16.

[142] Kanungo, P. K., Adhya, T. K., & Rao, V. R. (1995). Influence of repeated applications of carbofuran on nitrogenase activity and nitrogen-fixing bacteria associated with rhizosphere of tropical rice. *Chemosphere*, 31(5), 3249-3257.

[143] Karnatak, A. K., & Thorat, P. V. (2006). Effect of insecticidal micro-environment on the honey bee, *Apis mellifera* in *Brassica napus*. *J. Appl. Biosci.*, 32(1), 93-94.

[144] Kennedy, P. J., Conrad, K. F., Perry, J. N., Powell, D., Aegerter, J., Todd, A. D., Walters, K. F. A., & Powell, W. (2001). Comparison of two field-scale approaches for the study of effects of insecticides on polyphagous predators in cereals. *Appl. Soil Ecol.*, 17(3), 253-266.

[145] Kitulagodage, M., Buttemer, W., & Astheimer, L. (2011). Adverse effects of fipronil on avian reproduction and development: maternal transfer of fipronil to eggs in zebra finch Taeniopygia guttata and in ovo exposure in chickens *Gallus domesticus*. *Ecotoxicology*, 20(4), 653-660.

[146] Kjaer, C., Elmegaard, N., Axelsen, J. A., Andersen, P. N., & Seidelin, N. (1998). The impact of phenology, exposure and instar susceptibility on insecticide effects on a chrysomelid beetle population. *Pestic. Sci.*, 52(4), 361-371.

[147] Kobori, Y., & Amano, H. (2004). Effects of agrochemicals on life-history parameters of *Aphidius gifuensis* Ashmead (Hymenoptera: Braconidae). *Appl. Entomol. Zool.*, 39(2), 255-261.

[148] Koehler, H. H. (1997). Mesostigmata (Gamasina, Uropodina), efficient predators in agroecosystems. *Agric. Ecosyst. Environ.*, 62(2-3), 105-117.

[149] Koppenhofer, A. M., Cowles, R. S., Cowles, E. A., Fuzy, E. M., & Kaya, H. K. (2003). Effect of neonicotinoid synergists on entomopathogenic nematode fitness. *Entomol. exp. appl.*, 106(1), 7-18.

[150] Kreutzweiser, D. P., Good, K. P., Chartrand, D. T., Scarr, T. A., Holmes, S. B., & Thompson, D. G. (2008). Effects on litter-dwelling earthworms and microbial decomposition of soil-applied imidacloprid for control of wood-boring insects. *Pest Manage. Sci.*, 64(2), 112-118.

[151] Krokene, P. (1993). The effect of an insect growth regulator on grasshoppers (Acrididae) and non-target arthropods in Mali. *J. Appl. Entomol.*, 116(3), 248-266.

[152] Krupke, C. H., Hunt, G. J., Eitzer, B. D., Andino, G., & Given, K. (2012). Multiple routes of pesticide exposure for honey bees living near agricultural fields. *PLoS One*, 7(1), e29268.

[153] Ku, T. Y., & Wang, S. C. (1981). Insecticidal resistance of the major insect rice pests, and the effect of insecticides on natural enemies and non-target animals. *NTU Phytopathologist and Entomologist* [8], 1-18.

[154] Kumar, B. V., Boomathi, N., Kumaran, N., & Kuttalam, S. (2010). Non target effect of ethiprole+imidacloprid 80 WG on predators of rice planthoppers. *Madras Agric. J.,* 97(4/6), 153-156.

[155] Kumaran, N., Kumar, B. V., Boomathi, N., Kuttalam, S., & Gunasekaran, K. (2009). Non-target effect of ethiprole 10 SC to predators of rice planthoppers. *Madras Agric. J.,* 96(1/6), 208-212.

[156] Kunkel, B. A., Held, D. W., & Potter, D. A. (2001). Lethal and sublethal effects of bendiocarb, halofenozide, and imidacloprid on *Harpalus pennsylvanicus* (Coleoptera: Carabidae) following different modes of exposure in turfgrass. *J. Econ. Entomol.,* 94(1), 60-67.

[157] Kwon, Y. K., Wee, S. H., & Kim, J. H. (2004). Pesticide poisoning events in wild birds in Korea from 1998 to 2002. *J. Wildl. Dis.,* 40(4), 737-740.

[158] Lahr, J. (1998). An ecological assessment of the hazard of eight insecticides used in desert locust control, to invertebrates in temporary ponds in the Sahel. *Aquat.Ecol.,* 32(2), 153-162.

[159] Lakshmi, V. J., Krishnaiah, N. V., & Katti, G. R. (2010). Potential toxicity of selected insecticides to rice leafhoppers and planthoppers and their important natural enemies. *J. Biol. Control,* 24(3), 244-252.

[160] Langer-Jaesrich, M., Kohler, H. R., & Gerhardt, A. (2010). Assessing toxicity of the insecticide thiacloprid on *Chironomus riparius* (Insecta: Diptera) using multiple end points. *Arch. Environ. Contam. Toxicol.,* 58(4), 963-972.

[161] Lannacone, J., Onofre, R., & Huanquj, O. (2007). Ecotoxicological effects of cartap on *Poecilia reticulata* "guppy"(Poecilidae) and *Paracheirodon innesi* "Neon Tetra" (Characidae). *Gayana,* 71(2), 170-177.

[162] Lanzoni, A., Sangiorgi, L., Luigi, V.d., Consolini, L., Pasqualini, E., & Burgio, G. (2012). Evaluation of chronic toxicity of four neonicotinoids to *Adalia bipunctata L.* (Coleoptera: Coccinellidae) using a demographic approach. *IOBC/WPRS Bulletin,* 74, 211-217.

[163] Lao, W., Tsukada, D., Greenstein, D. J., Bay, S. M., & Maruya, K. A. (2010). Analysis, occurrence, and toxic potential of pyrethroids, and fipronil in sediments from an urban estuary. *Environ. Toxicol. Chem.,* 29(4), 843-851.

[164] Lee, S. J., Tomizawa, M., & Casida, J. E. (2003). Nereistoxin and cartap neurotoxicity attributable to direct block of the insect nicotinic receptor/channel. *J. Agric. Food Chem.,* 51(9), 2646-2652.

[165] Lewis, J. L., & Forschler, B. T. (2010). Impact of five commercial baits containing chitin synthesis inhibitors on the protist community in *Reticulitermes flavipes* (Isoptera: Rhinotermitidae). *Environ. Entomol.,* 39(1), 98-104.

[166] Li, X., Bao, C., Yang, D., Zheng, M., Li, X., & Tao, S. (2010). Toxicities of fipronil enantiomers to the honeybee *Apis mellifera* L. and enantiomeric compositions of fipronil in honey plant flowers. *Environ. Toxicol. Chem.*, 29(1), 127-132.

[167] Liess, M., & Beketov, M. (2011). Traits and stress: keys to identify community effects of low levels of toxicants in test systems. *Ecotoxicology*, 20(6), 1328-1340.

[168] Lim, R. P., & Wong, M. C. (1986). The effect of pesticides on the population dynamics and production of *Stenocypris major* Bairo (Ostracoda) in ricefieds. *Arch. Hydrobiol.*, 106(3), 421-427.

[169] Lima, Junior., I.d, S.d., Nogueira, R. F., Bertoncello, T. F., Melo, E., P.d, , Suekane, R., & Degrande, P. E. (2010). Seletividade de inseticidas sobre o complexo de predadores das pragas do algodoeiro. *Pesquisa Agropecuaria Tropical*, 40(3), 347-353.

[170] Linder, G., & Richmond, M. E. (1990). Feed aversion in small mammals as a potential source of hazard reduction for environmental chemicals: agrochemical case studies. *Environ. Toxicol. Chem.*, 9(1), 95-105.

[171] Lisker, E., Ensminger, M., Gill, S., & Goh, K. (2011). Detections of eleven organophosphorus insecticides and one herbicide threatening Pacific salmonids, *Oncorhynchus* spp., in California, 1991-2010. *Bull. Environ. Contam. Toxicol.*, 87(4), 355-360.

[172] Liu, Z., Dai, Y., Huang, G., Gu, Y., Ni, J., Wei, H., & Yuan, S. (2011). Soil microbial degradation of neonicotinoid insecticides imidacloprid, acetamiprid, thiacloprid and imidaclothiz and its effect on the persistence of bioefficacy against horsebean aphid *Aphis craccivora* Koch after soil application. *Pest Manage. Sci.*, 67(10), 1245-1252.

[173] Loureiro, S., Svendsen, C., Ferreira, A. L. G., Pinheiro, C., Ribeiro, F., & Soares, A. M. V. M. (2010). Toxicity of three binary mixtures to *Daphnia magna*: Comparing chemical modes of action and deviations from conceptual models. *Environ. Toxicol. Chem.*, 29(8), 1716-1726.

[174] Lu, C., Warchol, K. M., & Callahan, R. A. (2012). In situ replication of honey bee colony collapse disorder. *Bull. Insectology*, 65(1), 99-106.

[175] Lue, L. P., Lewis, C. C., & Melchor, V. E. (1984). The effect of aldicarb on nematode population and its persistence in carrots, soil and hydroponic solution. *J. Environ. Sci. Health B*, 19(3), 343-354.

[176] Maccagnani, B., Ferrari, R., Zucchi, L., & Bariselli, M. (2008). Nei medicai dell'emilia-romagna: difendersi dalle cavallette, ma tutelare le api. *Informatore Agrario*, 64(25), 53-56.

[177] Magagula, C. N., & Samways, M. J. (2000). Effects of insect growth regulators on *Chilocorus nigritus* (Fabricius) (Coleoptera: Coccinellidae), a non-target natural enemy of citrus red scale, *Aonidiella aurantii* (Maskell) (Homoptera: Diaspididae), in southern Africa: evidence from laboratory and field trials. *African Entomol.*, 8(1), 47-56.

[178] Malinowski, H. (2006). Bioroznorodnosc a ochrona lasu przed szkodliwymi owada-
mi. *Progress in Plant Protection*, 46(1), 319-325.

[179] Marletto, F., Patetta, A., & Manino, A. (2003). Laboratory assessment of pesticide toxicity to bumble bees. *Bull. Insectology*, 56(1), 155-158.

[180] Martikainen, E., Haimi, J., & Ahtiainen, J. (1998). Effects of dimethoate and benomyl
on soil organisms and soil processes- a microcosm study. *Appl. Soil Ecol.*, 9(1-3),
381-387.

[181] Martins, G. L. M., Toscano, L. C., Tomquelski, G. V., & Maruyama, W. I. (2009). Inseticidas no controle de *Anticarsia gemmatalis* (Lepidoptera: Noctuidae) e impacto sobre
aranhas predadoras em soja. *Revista Brasileira de Ciencias Agrarias*, 4(2), 128-132.

[182] Matsumura, F. (1985). Toxicology of Pesticides. Plenum Press 0-306-41979-3, New
York, USA.

[183] Maul, J. D., Brennan, A. A., Harwood, A. D., & Lydy, M. J. (2008). Effect of sediment-
asociated pyrethroids, fipronil, and metabolites on *Chironomus tentans* growth rate,
body mass, condition index, immobilization, and survival. *Environ. Toxicol. Chem.*,
27(12), 2582-2590.

[184] Mayer, D. F., & Lunden, J. D. (1994). Effects of the adjuvant Sylgard 309 on the hazard of selected insecticides to honey bees. *Bee Sci.*, 3(3), 135-138.

[185] Mayer, D. F., Patten, K. D., Macfarlane, R. P., & Shanks, C. H. (1994). Differences between susceptibility of four pollinator species (Hymenoptera: Apoidea) to field
weathered insecticide residues. *Melanderia*, 50, 24-27.

[186] Medrzycki, P., Montanari, R., Bortolotti, L., Sabatini, A. G., Maini, S., & Porrini, C.
(2003). Effects of imidacloprid administered in sublethal doses on honey bee behaviour. Laboratory tests. *Bull. Insectology*, 56(1), 59-62.

[187] Meher, H. C., Gajbhiye, V. T., Singh, G., Kamra, A., & Chawla, G. (2010). Persistence
and nematicidal efficacy of carbosulfan, cadusafos, phorate, and triazophos in soil
and uptake by chickpea and tomato crops under tropical conditions. *J. Agric. Food
Chem.*, 58(3), 1815-1822.

[188] Mesléard, F., Garnero, S., Beck, N., & Rosecchi, E. (2005). Uselessness and indirect
negative effects of an insecticide on rice field invertebrates. *Comptes Rendus Biologies*,
328(10-11), 955-62.

[189] Mineau, P. (1988). Avian mortality in agroecosystems. I. The case against granule insecticides in Canada. In: *Field Methods for the Study of Environmental Effects of Pesticides, Greaves*, M.P. et al., editors, British Crop Protection Council, London, 3-12.

[190] Mineau, P., & Whiteside, M. (2006). Lethal risk to birds from insecticide use in the
United States- A spatial and temporal analysis. *Environ. Toxicol. Chem.*, 25(5),
1214-1222.

[191] Miranda, G. R. B., Raetano, C. G., Silva, E., Daam, M. A., & Cerejeira, M. J. (2011). Environmental fate of neonicotinoids and classification of their potential risks to hypogean, epygean, and surface water ecosystems in Brazil. *Hum. Ecol. Risk Assess.*, 17(4), 981-995.

[192] Moens, J., Tirry, L., & Clercq, P.d. (2012). Susceptibility of cocooned pupae and adults of the parasitoid *Microplitis mediator* to selected insecticides. *Phytoparasitica*, 40(1), 5-9.

[193] Mommaerts, V., Reynders, S., Boulet, J., Besard, L., Sterk, G., & Smagghe, G. (2010). Risk assessment for side-effects of neonicotinoids against bumblebees with and without impairing foraging behavior. *Ecotoxicology*, 19(1), 207-215.

[194] Morandin, L. A., & Winston, M. L. (2003). Effects of novel pesticides on bumble bee (Hymenoptera: Apidae) colony health and foraging ability. *Environ. Entomol.*, 32(3), 555-563.

[195] Moreby, S. J., Southway, S., Barker, A., & Holland, J. M. (2001). A comparison of the effect of new and established insecticides on nontarget invertebrates on winter wheat fields. *Environ. Toxicol. Chem.*, 20(10), 2243-2254.

[196] Moser, S. E., & Obrycki, J. J. (2009). Non-target effects of neonicotinoid seed treatments; mortality of coccinellid larvae related to zoophytophagy. *Biol. Control*, 51(3), 487-492.

[197] Moser, V. C., Mc Daniel, K. L., Phillips, P. M., & Lowit, A. B. (2010). Time-course, dose-response, and age comparative sensitivity of N-methyl carbamates in rats. *Toxicol. Sci.*, 114(1), 113-123.

[198] Mosleh, Y. Y., Paris-Palacios, S., Couderchet, M., & Vernet, G. (2003). Acute and sublethal effects of two insecticides on earthworms (*Lumbricus terrestris* L.) under laboratory conditions. *Environ. Toxicol.*, 18(1), 1-8.

[199] Moulton, C. A., Flemming, W. J., & Purnell, C. E. (1996). Effects of two cholinesterase-inhibiting pesticides on freshwater mussels. *Environ. Toxicol. Chem.*, 15(2), 131-137.

[200] Mullin, C. A., Frazier, M., Frazier, J. L., Ashcraft, S., Simonds, R., D.v, E., & Pettis, J. S. (2010). High levels of miticides and agrochemicals in North American apiaries: implications for honey bee health. *PLoS One*, 5(3), e9754.

[201] Nakahira, K., Kashitani, R., Tomoda, M., Kodama, R., Ito, K., Yamanaka, S., Momoshita, M., Arakawa, R., & Takagi, M. (2011). Systemic nicotinoid toxicity against the predatory mirid *Pilophorus typicus*: residual side effect and evidence for plant sucking. *J. Faculty Agric. Kyushu University*, 56(1), 53-55.

[202] Naveed, M., Salam, A., Saleem, M. A., Rafiq, M., & Hamza, A. (2010). Toxicity of thiamethoxam and imidacloprid as seed treatments to parasitoids associated to control Bemisia tabaci. *Pakistan J. Zool.*, 42(5), 559-565.

[203] Norelius, E., & Lockwood, J. (1999). The effects of reduced agent-area insecticide treatments for rangeland grasshopper (Orthoptera: Acrididae) control on bird densities. *Arch. Environ. Contam. Toxicol.*, 37(4), 519-528.

[204] Ohnesorg, W. J., Johnson, K. D., & O'Neal, M. E. (2009). Impact of reduced-risk insecticides on soybean aphid and associated natural enemies. *J. Econ. Entomol.*, 102(5), 1816-1826.

[205] Oldroyd, B. P. (2007). What's killing American honey bees? *PLOS Biology*, 5(6), e168.

[206] Papachristos, D. P., & Milonas, P. G. (2008). Adverse effects of soil applied insecticides on the predatory coccinellid *Hippodamia undecimnotata* (Coleoptera: Coccinellidae). *Biol. Control*, 47(1), 77-81.

[207] Parker, M., & Goldstein, M. (2000). Differential toxicities of organophosphate and carbamate insecticides in the nestling European starling (*Sturnus vulgaris*). *Arch. Environ. Contam. Toxicol.*, 39(2), 233-242.

[208] Patterson, K. (1991). Killing the birds and the bees. *Environmental Action*, 23(1), 7-8.

[209] Peck, D. C. (2009). Comparative impacts of white grub (Coleoptera: Scarabaeidae) control products on the abundance of non-target soil-active arthropods in turfgrass. *Pedobiologia*, 52(5), 287-299.

[210] Peck, D. C., & Olmstead, D. (2010). Neonicotinoid insecticides disrupt predation on the eggs of turf-infesting scarab beetles. *Bull. Entomol. Res.*, 100(6), 689-700.

[211] Pekár, S. (1999). Foraging mode: a factor affecting the susceptibility of spiders (Araneae) to insecticide applications. *Pestic. Sci.*, 55(11), 1077-1082.

[212] Perfecto, I. (1990). Indirect and direct effects in a tropical agroecosystem: the maize-pest-ant system in Nicaragua. *Ecology*, 71(6), 2125-2134.

[213] Pettis, J., van Engelsdorp, D., Johnson, J., & Dively, G. (2012). Pesticide exposure in honey bees results in increased levels of the gut pathogen *Nosema*. *Naturwissenschaften*, 99(2), 153-158.

[214] Peveling, R., Mc William, A. N., Nagel, P., Rasolomanana, H., Raholijaona, Rakotomianina. L., Ravoninjatovo, A., Dewhurst, C. F., Gibson, G., Rafanomezana, S. , et al. (2003). Impact of locust control on harvester termites and endemic vertebrate predators in Madagascar. *J. Appl. Ecol.*, 40(4), 729-741.

[215] Porter, W. P., Green, S. M., Debbink, N. L., & Carlson, I. (1993). Groundwater pesticides: interactive effects of low concentrations of carbamates aldicarb and methomyl and the triazine metribuzin of thyroxine and somatotropin levels in white rats. *J. Toxicol. Environ. Health*, 40(1), 15-34.

[216] Potter, D. A., Buxton, M. C., Redmond, C. T., Patterson, C. G., & Powelu, A. J. (1990). Toxicity of pesticides to earthworms (Oligochaeta: Lumbricidae) and effect on thatch degradation in Kentucky bluegrass turf. *J. Econ. Entomol.*, 83(6), 2362-2369.

[217] Potts, G. R. (1986). The Partridge - Pesticides, Predation and Conservation. *Collins, London, UK.*

[218] Pozzebon, A., Duso, C., Tirello, P., & Ortiz, P. B. (2011). Toxicity of thiamethoxam to *Tetranychus urticae* Koch and *Phytoseiulus persimilis* Athias-Henriot (Acari Tetranychidae, Phytoseiidae) through different routes of exposure. *Pest Manage. Sci.*, 67(3), 352-359.

[219] Prabhaker, N., Castle, S., Byrne, F., Henneberry, T. J., & Toscano, N. C. (2006). Establishment of baseline susceptibility data to various insecticides for Homalodisca coagulata (Homoptera: Cicadellidae) by comparative bioassay techniques. *J. Econ. Entomol.*, 99(1), 141-154.

[220] Preetha, G., Manoharan, T., Stanley, J., & Kuttalam, S. (2010a). Impact of chloronicotinyl insecticide, imidacloprid on egg, egg-larval and larval parasitoids under laboratory conditions. *J. Plant Protection Res.*, 50(4), 535-540.

[221] Preetha, G., Stanley, J., Suresh, S., & Samiyappan, R. (2010b). Risk assessment of insecticides used in rice on miridbug, *Cyrtorhinus lividipennis* Reuter, the important predator of brown planthopper, *Nilaparvata lugens* (Stal.). *Chemosphere*, 80(5), 498-503.

[222] Printes, L. B., & Callaghan, A. (2004). A comparative study on the relationship between acetylcholinesterase activity and acute toxicity in *Daphnia magna* exposed to acetylcholineesterase insecticides. *Environ. Toxicol. Chem.*, 23(5), 1241-1247.

[223] Qin, G. Q., Presley, S. M., Anderson, T. A., Gao, W. M., & Maul, J. D. (2011). Effects of predator cues on pesticide toxicity: toward an understanding of the mechanism of the interaction. *Environ. Toxicol. Chem.*, 30(8), 1926-1934.

[224] Rainwater, T. R., Leopold, V. A., Hooper, M. J., & Kendall, R. J. (1995). Avian exposure to organophosphorous and carbamate pesticides on a coastal South Carolina golf course. *Environ. Toxicol. Chem.*, 14(12), 2155-2161.

[225] Ram, S., & Gupta, G. P. (1994). Bioefficacy of systemic insecticides against target pest jassid (*Amrasca devastans*) and their impact on non-target soil microarthropods in cotton (*Gossypium* spp.). *Indian J. Entomol.*, 56(4), 313-321.

[226] Redoan, A. C., Carvalho, G. A., Cruz, I., Figueiredo, M.d. L. C., & Silva, R. B.d. (2010). Efeito de inseticidas usados na cultura do milho (*Zea mays* L.) sobre ninfas e adultos de *Doru luteipes* (Scudder) (Dermaptera: Forficulidae) em semicampo. *Revista Brasileira de Milho e Sorgo*, 9(3), 223-235.

[227] Rodrigues, E.d. L., & Fanta, E. (1998). Liver histopathology of the fish *Brachydanio rerio* Hamilton-Buchman after acute exposure to sublethal levels of the organophosphate dimethoate 500. *Revista Brasileira de Zoologia*, 15(2), 441-450.

[228] Rortais, A., Arnold, G., Halm, M. P., & Touffet-Briens, F. (2005). Modes of honeybees exposure to systemic insecticides: estimated amounts of contaminated pollen and nectar consumed by different categories of bees. *Apidologie*, 36(1), 71-83.

[229] Rudolph, S. G., Zinkl, J. G., Anderson, D. W., & Shea, P. J. (1984). Prey-capturing ability of American kestrels fed DDE and acephate or acephate alone. *Arch. Environ. Contam. Toxicol.*, 13, 367-372.

[230] Rust, M. K., & Saran, R. K. (2008). Toxicity, repellency, and effects of acetamiprid on western subterranean termite (Isoptera: Rhinotermitidae). *J. Econ. Entomol.*, 101(4), 1360-1366.

[231] Saber, M. (2011). Acute and population level toxicity of imidacloprid and fenpyroximate on an important egg parasitoid, *Trichogramma cacoeciae* (Hymenoptera: Trichogrammatidae). *Ecotoxicology*, 20(6), 1476-1484.

[232] San, Miguel. A., Raveton, M., Lemperiere, G., & Ravanel, P. (2008). Phenylpyrazoles impact on *Folsomia candida* (Collembola). Soil Biol. *Biochem.*, 40(9), 2351-2357.

[233] Sánchez, Meza. J. C., Avila, Perez. P., Borja, Salin. M., Pacheco, Salazar. V. F., & Lapoint, T. (2010). Inhibition of cholinesterase activity by soil extracts and predicted environmental concentrations (PEC) to select relevant pesticides in polluted soils. *J. Environ. Sci. Health B*, 45(3), 214-221.

[234] Sánchez-Bayo, F. (2009). From simple toxicological models to prediction of toxic effects in time. *Ecotoxicology*, 18(3), 343-354.

[235] Sánchez-Bayo, F. (2012). Insecticides mode of action in relation to their toxicity to non-target organisms. *J. Environ. Anal. Toxicol.*, S4, S4-002.

[236] Sánchez-Bayo, F., & Goka, K. (2005). Unexpected effects of zinc pyrithione and imidacloprid on Japanese medaka fish (*Oryzias latipes*). *Aquat. Toxicol.*, 74(4), 285-293.

[237] Sánchez-Bayo, F., & Goka, K. (2006). Ecological effects of the insecticide imidacloprid and a pollutant from antidandruff shampoo in experimental rice fields. *Environ. Toxicol. Chem.*, 25(6), 1677-1687.

[238] Sánchez-Bayo, F., Yamashita, H., Osaka, R., Yoneda, M., & Goka, K. (2007). Ecological effects of imidacloprid on arthropod communities in and around a vegetable crop. *J. Environ. Sci. Health B*, 42(3), 279-286.

[239] Schaefer, C. H., Miura, T., Dupras, E. F. Jr, Wilder, W. H., & Mulligan, F. S. III. (1988). Efficacy of CME 134 against mosquitoes (Diptera: Culicidae): effects on nontarget organisms and evaluation of potential chemical persistence. *J. Econ. Entomol.*, 81(4), 1128-1132.

[240] Schmuck, R. (2004). Effects of a chronic dietary exposure of the honeybee *Apis mellifera* (Hymenoptera: Apidae) to imidacloprid. *Arch. Environ. Contam. Toxicol.*, 47(4), 471-478.

[241] Schmuck, R., Schöning, R., Stork, A., & Schramel, O. (2001). Risk posed to honeybees (*Apis mellifera* L, Hymenoptera) by an imidacloprid seed dressing of sunflowers. *Pest Manage. Sci.*, 57(3), 225-238.

[242] Schmuck, R., Stadler, T., & Schmidt, H. W. (2003). Field relevance of a synergistic effect observed in the laboratory between an EBI fungicide and a chloronicotinyl insecticide in the honeybee (*Apis mellifera* L, Hymenoptera). *Pest Manage. Sci.*, 59(3), 279-286.

[243] Schneider, F. (1966). Some pesticide-wildlife problems in Switzerland. *J. Appl. Ecol.*, 3, 15-20.

[244] Sechser, B., & Freuler, J. (2003). The impact of thiamethoxam on bumble bee broods (*Bombus terrestris* L.) following drip application in covered tomato crops. *Anzeiger fur Schadlingskunde*, 76(3), 74-77.

[245] Sétamou, M., Rodriguez, D., Saldana, R., Schwarzlose, G., Palrang, D., & Nelson, S. D. (2010). Efficacy and uptake of soil-applied imidacloprid in the control of Asian citrus psyllid and a citrus leafminer, two foliar-feeding citrus pests. *J. Econ. Entomol.*, 103(5), 1711-9.

[246] Sharma, D. R. (2010). Bioefficacy of insecticides against peach leaf curl aphid, *Brachycaudus helichrysi* (Kaltenbach) in Punjab. *Indian J. Entomol.*, 72(3), 217-222.

[247] Shi, X., Jiang, L., Wang, H., Qiao, K., Wang, D., & Wang, K. (2011). Toxicities and sublethal effects of seven neonicotinoid insecticides on survival, growth and reproduction of imidacloprid-resistant cotton aphid, *Aphis gossypii*. *Pest Manage. Sci.*, 67(12), 1528-1533.

[248] Shimoda, T., Yara, K., & Kawazu, K. (2011). The effects of eight insecticides on the foraging behavior of the parasitoid wasp *Cotesia vestalis*. *J. Plant Interact.*, 6(2/3), 189-190.

[249] Simpson, I. C., Roger, P. A., Oficial, R., & Grant, I. F. (1994). Effects of nitrogen fertiliser and pesticide management on floodwater ecology in a wetland ricefield. *Biol. Fertil. Soils*, 18(3), 219-227.

[250] Smelt, J. H., Crum, S. J. H., Teunissen, W., & Leistra, M. (1987). Accelerated transformation of aldicarb, oxamyl and ethoprophos after repeated soil treatments. *Crop Protection*, 6(5), 295-303.

[251] Smith, S. F., & Krischik, V. A. (1999). Effects of systemic imidacloprid on *Coleomegilla maculata* (Coleoptera: Coccinellidae). *Environ. Entomol.*, 28(6), 1189-1195.

[252] Sokolov, I. M. (2000). How does insecticidal control of grasshoppers affect non-target arthropods?

[253] Song, M. Y., & Brown, J. J. (2006). Influence of fluctuating salinity on insecticide tolerance of two euryhaline arthropods. *J. Econ. Entomol.*, 99(3), 745-751.

[254] Sprague, L. A., & Nowell, L. H. (2008). Comparison of pesticide concentrations in streams at low flow in six metropolitan areas of the United States. *Environ. Toxicol. Chem.*, 27(2), 288-298.

[255] Srinivas, K., & Madhumathi, T. (2005). Effect of insecticide applications on the preda-
       tor population in rice ecosystem of Andhra Pradesh. *Pest Manage. Econ. Zool.*, 13(1),
       71-75.

[256] Stadler, T., Martinez, Gines. D., & Buteler, M. (2003). Long-term toxicity assessment
       of imidacloprid to evaluate side effects on honey bees exposed to treated sunflower
       in Argentina. *Bull. Insectology*, 56(1), 77-81.

[257] Stapel, J. O., Cortesero, A. M., & Lewis, W. J. (2000). Disruptive sublethal effects of
       insecticides on biological control: altered foraging ability and life span of a parasitoid
       after feeding on extrafloral nectar of cotton treated with systemic insecticides. *Biol.
       Control*, 17, 243-249.

[258] Stara, J., Ourednickova, J., & Kocourek, F. (2011). Laboratory evaluation of the side
       effects of insecticides on *Aphidius colemani* (Hymenoptera: Aphidiidae), *Aphidoletes
       aphidimyza* (Diptera: Cecidomyiidae), and *Neoseiulus cucumeris* (Acari: Phytoseidae). *J.
       Pest Sci.*, 84(1), 25-31.

[259] Stark, J., & Vargas, R. (2005). Toxicity and hazard assessment of fipronil to *Daphnia
       pulex Ecotoxicol. Environ. Saf.* , 62(1), 11-16.

[260] Stark, J. D., Banks, J. E., & Vargas, R. (2004). How risky Is risk assessment: the role
       that life history strategies play in susceptibility of species to stress. *PNAS*, 101(3),
       732-736.

[261] Starner, K., & Goh, K. (2012). Detections of the neonicotinoid insecticide imidacloprid
       in surface waters of three agricultural regions of California, USA, 2010-2011. *Bull. En-
       viron. Contam. Toxicol.*, 88(3), 316-321.

[262] Steinbauer, M. J., & Peveling, R. (2011). The impact of the locust control insecticide
       fipronil on termites and ants in two contrasting habitats in northern Australia. *Crop
       Protection*, 30(7), 814-825.

[263] Sterk, G., & Benuzzi, M. (2004). Nuovi fitofarmaci, prove di tossicita sui bombi in ser-
       ra. *Colture Protette*, 33(1), 75-77.

[264] Stevens, M. M., Burdett, A. S., Mudford, E. M., Helliwell, S., & Doran, G. (2011). The
       acute toxicity of fipronil to two non-target invertebrates associated with mosquito
       breeding sites in Australia. *Acta Tropica*, 117(2), 125-130.

[265] Stoughton, Sarah. J., Karsten, Liber., Joseph, Culp., & Allan, Cessna. (2008). Acute
       and chronic toxicity of imidacloprid to the aquatic invertebrates *Chironomus tentans*
       and *Hyalella azteca* under constant- and pulse-exposure conditions. *Arch. Environ.
       Contam. Toxicol.*, 54(4), 662-673.

[266] Suchail, S., Guez, D., & Belzunces, L. P. (2001). Discrepancy between acute and
       chronic toxicity induced by imidacloprid and its metabolites in *Apis mellifera. Environ.
       Toxicol. Chem.*, 20(11), 2482-2486.

[267] Sugiyama, K., Katayama, H., & Saito, T. (2011). Effect of insecticides on the mortalities of three whitefly parasitoid species, *Eretmocerus mundus*, *Eretmocerus eremicus* and *Encarsia formosa* (Hymenoptera: Aphelinidae). *Appl. Entomol. Zool.*, 46(3), 311-317.

[268] Szczepaniec, A., Creary, S. F., Laskowski, K. L., Nyrop, J. P., & Raupp, M. J. (2011). Neonicotinoid insecticide imidacloprid causes outbreaks of spider mites on elm trees in urban landscapes. *PLoS One*, 6(5), e20018.

[269] Szendrei, Z., Grafius, E., Byrne, A., & Ziegler, A. (2012). Resistance to neonicotinoid insecticides in field populations of the Colorado potato beetle (Coleoptera: Chrysomelidae). *Pest Manage. Sci.*, 68(6), 941-946.

[270] Takada, Y., Kawamura, S., & Tanaka, T. (2001). Effects of various insecticides on the development of the egg parasitoid *Trichogramma dendrolimi* (Hymenoptera: Trichogrammatidae). *J. Econ. Entomol.*, 94(6), 1340-1343.

[271] Tanaka, T., & Minakuchi, C. (2011). Insecticides and parasitoids. *In: Insecticides- Advances in Integrated Pest Management* , Perveen, F., editor, InTech, Rijeka, Croatia, 115-140.

[272] Tapparo, A., Giorio, C., Marzaro, M., Marton, D., Solda, L., & Girolami, V. (2011). Rapid analysis of neonicotinoid insecticides in guttation drops of corn seedlings obtained from coated seeds. *J. Environ. Monit.*, 13(6), 1564-1568.

[273] Teeters, B. S., Johnson, R. M., Ellis, M. D., & Siegfried, B. D. (2012). Using video-tracking to assess sublethal effects of pesticides on honey bees (*Apis mellifera* L.). *Environ. Toxicol. Chem.*, 31(6), 1349-1354.

[274] Tennekes, H. A. (2010a). The significance of the Druckrey-Küpfmüller equation for risk assessment - The toxicity of neonicotinoid insecticides to arthropods is reinforced by exposure time. *Toxicology*, 276(1), 1-4.

[275] Tennekes, H. A. (2010b). The Systemic Insecticides: A Disaster in the Making. ETS Nederland BV, Zutphen, The Netherlands. 978-90-79627-06-6

[276] Tennekes, H. A., & Sánchez-Bayo, F. (2012). Time-dependent toxicity of neonicotinoids and other toxicants: implications for a new approach to risk assessment. *J. Environ. Anal. Toxicol.*, S4, S4-001.

[277] Thacker, J. R. M., & Jepson, P. C. (1993). Pesticide risk assessment and non-target invertebrates: integrating population depletion, population recovery, and experimental design. *Bull. Environ. Contam. Toxicol.*, 51(4), 523-531.

[278] Thomazoni, D., Soria, M. F., Kodama, C., Carbonari, V., Fortunato, R. P., Degrande, P. E., & Valter, Junior. V. A. (2009). Selectivity of insecticides for adult workers of *Apis mellifera* (Hymenoptera: Apidae). *Revista Colombiana de Entomologia*, 35(2), 173-176.

[279] Thompson, H., Walker, C., & Hardy, A. (1991). Changes in activity of avian serum esterases following exposure to organophosphorus insecticides. *Arch. Environ. Contam. Toxicol.*, 20(4), 514-518.

[280] Thornton, M., Miller, J., Hutchinson, P., & Alvarez, J. (2010). Response of potatoes to soil-applied insecticides, fungicides, and herbicides. *Potato Research*, 53(4), 351-358.

[281] Tian, H., Ru, S., Bing, X., & Wang, W. (2010). Effects of monocrotophos on the reproductive axis in the male goldfish (*Carassius auratus*): potential mechanisms underlying vitellogenin induction. *Aquat. Toxicol.*, 98(1), 67-73.

[282] Tian, J., Chen, Y., Li, Z., Peng, Y., & Ye, G. (2011). Assessment of effect of transgenic rice with cry1Ab gene and two insecticides on immune of non-target natural enemy, *Pardosa pseudoannulata. Chinese J. Biol. Control* , 27(4), 559-563.

[283] Tomizawa, M., Lee, D. L., & Casida, J. E. (2000). Neonicotinoid insecticides: molecular features conferring selectivity for insect versus mammalian nicotinic receptors. *J. Agric. Food Chem.*, 48(12), 6016-6024.

[284] Tomlin, C. D. S. (2009). The e-Pesticide Manual. Tomlin, C.D.S.,editor. 12 ed. Surrey, U.K.: British Crop Protection Council.

[285] Tremolada, P., Mazzoleni, M., Saliu, F., Colombo, M., & Vighi, M. (2010). Field trial for evaluating the effects on honeybees of corn sown using cruiser® and Celest ®Treated Seeds. *Bull. Environ. Contam. Toxicol.*, 85(3), 229-234.

[286] Turner, A. S., Bale, J. S., & Clements, R. O. (1987). The effect of a range of pesticides on non-target organisms in the grassland environment. Proceedings, Crop Protection in Northern Britain '87, Dundee University, 15-17 March., 290-295.

[287] van Dijk, T. C. (2010). Effects of neonicotinoid pesticide pollution of Dutch surface water on non-target species abundance. *Universiteit Utrecht.*, 77.

[288] Van Gestel, C. A. M. (1992). Validation of earthworm toxicity tests by comparison with field studies: a review of benomyl, carbendazim, carbofuran, and carbaryl. *Ecotoxicol. Environ. Saf.*, 23, 221-236.

[289] van Timmeren, S., Wise, J. C., & Isaacs, R. (2012). Soil application of neonicotinoid insecticides for control of insect pests in wine grape vineyards. *Pest Manage. Sci.*, 68(4), 537-542.

[290] Vasuki, V. (1992). Sublethal effects of hexaflumuron, an insect growth regulator, on some non-target larvivorous fishes. *Indian J. Exp. Biol.* , 30(12), 1163-1165.

[291] Verghese, A. (1998). Effect of imidacloprid on mango hoppers, *Idioscopus* spp. (Homoptera: Cicadellidae). *Pest Manage. Hort. Ecosyst.*, 4(2), 70-74.

[292] Vernon, R. S., Herk, W.v., Tolman, J., Saavedra, H. O., Clodius, M., & Gage, B. (2008). Transitional sublethal and lethal effects of insecticides after dermal exposures to five economic species of wireworms (Coleoptera: Elateridae). *J. Econ. Entomol.*, 101(2), 365-374.

[293] Vidau, C., Diogon, M., Aufauvre, J., Fontbonne, R., Vigues, B., Brunet, J. L., Texier, C., Biron, D. G., Blot, N., El -Alaoui, H., et al. (2011). Exposure to sublethal doses of fipronil and thiacloprid highly increases mortality of honeybees previously infected by *Nosema ceranae*. *PLoS One(June)*, e21550.

[294] Vogt, H., Just, J., & Grutzmacher, A. (2009). Einfluss von Insektiziden im Obstbau auf den Ohrwurm *Forficula auricularia*. *Mitteilungen der Deutschen Gesellschaft fur allgemeine und angewandte Entomologie*, 17, 211-214.

[295] Vyas, N. B., Kuenzel, W. J., Hill, E. F., & Sauer, J. R. (1995). Acephate affects migratory orientation of the white-throated sparrow (*Zonotrichia albicollis*). *Environ. Toxicol. Chem.*, 14(11), 1961-1965.

[296] Wada, S., & Toyota, K. (2008). Effect of three organophosphorous nematicides on non-target nematodes and soil microbial community. *Microbes and Environments*, 23(4), 331-336.

[297] Walker, C. H., Hopkin, S. P., Sibly, R. M., & Peakall, D. B. (2001). Principles of Ecotoxicology. 2nd), Taylor & Francis, Glasgow, UK.0-7484-0940-8

[298] Walker, M. K., Stufkens, M. A. W., & Wallace, A. R. (2007). Indirect non-target effects of insecticides on Tasmanian brown lacewing (*Micromus tasmaniae*) from feeding on lettuce aphid (*Nasonovia ribisnigri*). *Biol. Control*, 43(1), 31-40.

[299] Wang, Y., Yu, R., Zhao, X., An, X., Chen, L., Wu, C., & Wang, Q. (2012). Acute toxicity and safety evaluation of neonicotinoids and macrocyclic lactiones to adult wasps of four *Trichogramma* species (Hymenoptera: Trichogrammidae). *Acta Entomologica Sinica*, 55(1), 36-45.

[300] Way, M. J., & Scopes, N. E. A. (1968). Studies on the persistence and effects on soil fauna of some soil-applied systemic insecticides. *Ann. Appl. Biol.*, 62, 199-214.

[301] Webster, T. C., & Peng, Y. S. (1989). Short-term and long-term effects of methamidophos on brood rearing in honey bee (Hymenoptera: Apidae) colonies. *J. Econ. Entomol.*, 82(1), 69-74.

[302] White, D. H., & Segmak, J. T. (1990). Brain cholinesterase inhibition in songbirds from pecan groves sprayed with phosalone and disulfoton. *J. Wildl. Dis.*, 26, 103-106.

[303] Wirth, E. F., Pennington, P. L., Lawton, J. C., De Lorenzo, M. E., Bearden, D., Shaddrix, B., Sivertsen, S., & Fulton, M. H. (2004). The effects of the contemporary-use insecticide (fipronil) in an estuarine mesocosm. *Environ. Pollut.*, 131(3), 365-371.

[304] Yang, E. C., Chuang, Y. C., Chen, Y. L., & Chang, L. H. (2008). Abnormal foraging behavior induced by sublethal dosage of imidacloprid in the honey bee (Hymenoptera: Apidae). *J. Econ. Entomol.*, 101(6), 1743-1748.

[305] Yen, J. H., Lin, K. H., & Wang, Y. S. (2000). Potential of the insecticides acephate and methamidophos to contaminate groundwater. *Ecotoxicol. Environ. Saf.*, 45(1), 79-86.

[306] Zang, Y., Zhong, Y., Luo, Y., & Kong, Z. M. (2000). Genotoxicity of two novel pesti-
       cides for the earthworm, *Eisenia fetida*. *Environ. Pollut.*, 108(2), 271-278.

[307] Zhang, A., Kaiser, H., Maienfisch, P., & Casida, J. E. (2000). Insect nicotinic acetylcho-
       line receptor: conserved neonicotinoid specificity of [3H]imidacloprid binding site. *J.
       Neurochem.*, 75(3), 1294-1303.

[308] Zhou, S., Dong, Q., Li, S., Guo, J., Wang, X., & Zhu, G. (2009). Developmental toxicity
       of cartap on zebrafish embryos. *Aquat. Toxicol.*, 95(4), 339-346.

Insecticides Against Pests of Urban Area, Forests and Farm Animals

Chapter 6

Bait Evaluation Methods for Urban Pest Management

Bennett W. Jordan, Barbara E. Bayer,
Philip G. Koehler and Roberto M. Pereira

Additional information is available at the end of the chapter

1. Introduction

Baits are a preferred type of formulation used in urban pest management, especially for the control of cockroaches, ants, and increasingly termites. With precise placement in areas away from contact with human population, especially children, and a reduced rate of active ingredient (AI) application in a given structure area, baits are more economical and pose less risk for consumers and the environment than other formulations. However, baits are very difficult to evaluate for efficacy. For baits, pest acceptance and horizontal transfer of bait are essential in order to control pest populations.

Baits are composed of one or more insecticide active ingredients incorporated into an attractive food matrix, which varies according to the type of target pest, and even according to species within a certain pest type. Although commercial development of species-specific baits may represent a serious commercial problem due to limited market, this has been done in the past, for instance with imported fire ant baits in the USA. However, typically, baits are developed to target a group of similar insects, e.g., cockroaches, which may vary in their response to the bait formulation, resulting in varying degrees of control depending on the pest population composition.

In order to perform successfully, baits must attract the target insect and be ingested in sufficient amount that will cause the desirable level of control in the pest population. For non-social insects, such as cockroaches, transfer of the active ingredient among different segments of the pest population (e.g., adults and immature forms, reproductives, etc) is desirable but not necessarily an essential characteristic of the baits. However, in social insects (ants and termites) the transfer of the active ingredients between the foragers and the remaining of the population, and specially the reproductive caste, is essential in providing adequate control of the pest population within reasonable time.

2. Cockroach bait evaluations

Cockroach consumption of a bait and subsequent control can be complex. More than one of these cockroach pests may occur at a location with each having its own food requirements [1], susceptibility to insecticides [2], and aversion to certain bait formulations [3, 4]. Additionally, within each species, feeding patterns [5] and insecticide susceptibility [6] can vary among stadia.

When cockroach baits are placed close to harborage, they are usually in direct competition with other food and water resources. Therefore, baits, which are often gels with 40-60% moisture [7, 8], need to out-compete other sources of dry food as well as other moisture sources so cockroaches will consume them. Although cockroach control by baits is primarily due to bait consumption, not all insects within a population are actively seeking food, so not all individuals may consume a bait. Besides consumption, cockroaches are exposed to insecticidal active ingredients by contacting baits with antennae or palps. Contact exposure results in some toxicant transfer and mortality [9]. Also, cockroaches can be affected by contact with small amounts of translocated active ingredient (trampling), or when they consume contaminated feces (coprophagy), dead or dying cockroaches (cannibalism), or vomit (emetophagy) [10, 11, 12, 13]. These effects can result in secondary and sometimes even tertiary mortality [14].

The combination of consumption, contact, and secondary exposure results in mortality of various cockroach species and stadia (Figure 1).

Therefore, it is important to consider the consumption of insecticidal baits and the consequent ingestion of active ingredients, the mortality of different life stages within the pest cockroach population, and any possible effect of indirect to the active ingredient in the bait without actual bait consumption.

Consumption. In testing consumption in the laboratory, the use of mixed-age cockroach populations (adults of both sex and nymphs all in the same arena) is important to simulate bait consumption in natural infestations (Figure 2).

Weight-change controls, which are protected from consumption by the insects but otherwise under the same conditions as the bait being tested, must be used so adjustments can be made due to moisture change in bait and any other used in the experiments. To better understand potential differences between different products when consumed by different insect populations, it is important to estimate the consumption of bait in relation to the size of the insect consuming it. Bait consumption (B_{con}) per g of insect can calculated as using the following equation:

Consumption (mg)/g cockroach = $\left(\left(F_B - \left\{ F_B * \left[\left(WC_B - WC_A \right) / WC_B \right] \right\} - F_A \right) / W_t \right)$

where F_B is the weight of bait (mg) available to cockroaches at the start of the experiment, WC_B is the weight of weight-change control bait (mg) before the experiment, WC_A is the weight of weight-change control bait (mg) at the end of the experiment, F_A is the weight of bait (mg) remaining after the experiment and consumption by the cockroaches, and W_t is the total weight of cockroaches placed in the arena.

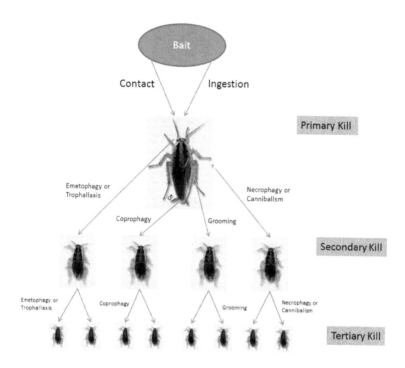

Figure 1. Cockroach baits can control populations in several ways, including primary kill, secondary kill, and tertiary kill.

Precise measurement of gel bait consumption by cockroaches is complicated by several factors:

a. Cockroaches are likely to remove and spread bait that is never eaten, especially in small arenas with large insect populations. Although the contact with the bait without consumption can be an important element in producing the total cockroach mortality, it can cause the measurement of bait consumption to be very unreliable and variable among the different experimental replicates. The use of larger arenas with smaller cockroach populations and plenty of harborage areas away from the bait can help in minimize this effect.

b. Rapid water loss cause gel bait especially, but also other bait forms, to change weight even in the absence of any consumptions. This factor is most severe on baits with very high water content, and formulations that do not limit water loss. Unless changes in water content can be estimated precisely, consumption can be overestimated. Any other calculations that result from the bait consumption (active ingredient consumption, bait preference in relation to alternative food, etc) can be greatly affected by over or under-estimation of the bait. Also, an estimate of the amount of bait necessary for control of a population of cockroaches will be greatly affected by any miscalculations due to imperfect water loss estimates. Rapid water loss can also affect the palatability and nutritional content of the baits, which will greatly affect the bait's effectiveness as well as any measurements associated with its consumption.

c.   Differential consumption of food by different cockroach life stadia can vary over short periods of time. For experimentation, strict selection of insects within specific age groups can minimize any problems associated with the inclusion of insects that will not consume any bait, or any other food, for some time into the experimental period.

One solution to minimize some of the effects on measurements of consumption is to limit the measurement to a specific time period (e.g., 24 h following the initial exposure). Although water loss and trampling may be more severe during the initial hours after application, some of these problems can be resolved by using weight-loss controls. Limitations on how the cockroaches can reach the bait may be used to limit trampling on the baits but researchers must be careful in not limiting access to the bait, especially if bait stations are crowded.

Figure 2. Cockroach arena set up for bait evaluations. Harborage and water vials are on left, bait and untreated food is in center, and protected water controls are on right.

Active Ingredient Consumption. Because mortality is dependent on the actual amount of active ingredient consumed by the insects, it is important to determine that consumption (Figure 3).

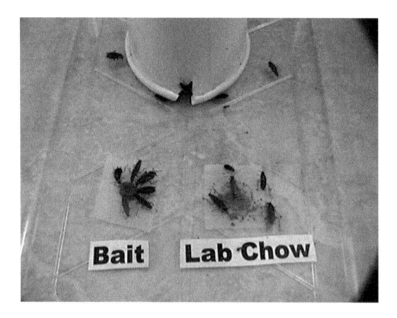

Figure 3. Cockroaches leave harborages in choice tests and can choose to ingest either bait or untreated food, like laboratory chow.

Baits with lower active ingredient content may be more palatable to the insect and be consumed in relatively high amounts, while a bait with high active ingredient content may partially deter consumption by the pest population. The active ingredient (AI) consumption/g cockroach can be calculated as follows:

AI / g cockroach (μg) = B_{con} * C_{ai} * 1000

where B_{con} is bait consumption (mg)/g cockroach, and C_{ai} is the percentage of active ingredient in the formulated bait.

From the point of view of pest control, the delivery of the active ingredients is the most critical factor when using baits. Different active ingredient concentrations combined with varying bait palatability determine the total amount of active ingredient ingested by the pests and their consequent mortality (Figure 4).

In general, when consuming baits, individual cockroaches will consume more active ingredient than necessary to cause death. Depending on the bait product and the cockroach species, the quantity of active ingredient ingested can vary from just above what is needed kill the insect to more than 1000 times the LD_{50}. This excess consumption of AI may be important in avoiding or delaying development of insecticide resistance by killing virtually all cockroaches exposed to these baits [15].

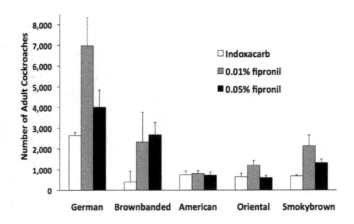

Figure 4. Number of adult cockroaches that could potentially consume and be killed by a 30-g tube of gel bait based on the consumption/g cockroach and the weights of average adult cockroaches from 5 different species (German: 56.7 ± 0.83 mg; brownbanded: 82.8 ± 0.49 mg, American: 959.4 ± 8.44 mg, smokeybrown 679.6 ± 19.02 mg, oriental 499.6 ± 9.51 mg).

Speed of kill is another aspect associated with the amount of active ingredient ingested by the cockroach. Although a quick kill can be advantageous in eliminating the pest problem within short time after the control action, other aspects, such as the possibility of transferring the active ingredient to other insects, can be maximized by limiting the amount of active ingredient that any individual cockroach will consume. Thus, understanding the different factors affecting the amount of active ingredient likely to be consumed by the average cockroach will help in the development of baits with greatest chances for success. Differences in susceptibility to insecticides [2, 6] and in feeding patterns [5] for different life stages of the pest, may also affect how fast bait materials will act. Adult males and non-gravid females are more likely to encounter and consume bait due to more consistent feeding patterns than nymphs.

Percent Bait Consumption. In the development of bait product, it is important that the bait competes well against other preferred food that the insect may find in their habitat. Thus, some measurement of bait preference over other foods is important in the development of baits. An important indication of how well a bait product will perform is the percentage of the total food consumption that is actually represented by the bait. The greater the percentage of bait in the total food consumption, the greater and faster mortality can be expected in the pest population. Percentage bait selection over an alternative preferred food can be determined by calculating the percentage of bait in the total food consumption, as follows:

% Bait Selection = $[B_{con}/ (B_{con} + AF_{con})] * 100$

where, B_{con} is the bait consumption (mg) and $+ AF_{con}$ is alternative food consumption (mg).

Cockroach baits have been optimized for some of the most common pest species, and minor pest species may be much less attracted to the commercial baits. Preference for the baits may be due to a balanced mixture of moisture and nutrients specially given the high water need by most cockroach species. However, different species of cockroaches have different water needs and this will be reflected in the percent of bait consumed when alternative foods vary widely in water content in relation to the bait. With high water losses [16], high cuticular permeability [17], high metabolic rate [17], cockroaches need to balance moisture and nutritional needs [18] in order to survive and reproduce at optimal levels.

Secondary Effects. Secondary effects to the consumption of bait may be very important, especially in relation to the portion of the cockroach population that may not consume any of the bait. Gravid females and nymphs during the ecdysis may not consume the bait before dries up or is consume by more aggressive bait consumers. The only way to affect the cockroaches that do not consume the bait directly is through secondary effects that result from consumption or other contact with bait contaminated debris, feces, and other materials.

To test these secondary effects, the arenas used in primary consumption studies should be set aside and remain unchanged except for the removal of live and dead cockroaches and any unused bait. Any remaining alternative food as well as harborages, water vials, containers, frass and any debris can be left in the arenas. These arenas can then be supplied with fresh food and water and a new population of cockroaches. Secondary effects (mortality) due to the contact with an environment contaminated by cockroaches consuming insecticidal baits can be evaluated against separate populations of cockroaches which are added to the contaminated arenas immediately after the primary consumption experiment and at different time intervals. A mixed population of cockroaches should also be used for these secondary effect experiments so that mortality in natural infestations can be simulated. Data similar to that collected in primary consumption experiments can be obtained in these experiments.

To reach the portion of the cockroach population that will not consume the bait, perhaps the best solution is the design of baits that offers greater opportunity for secondary mortality through contact with either dying insects or debris moved by insect that visit and consume the baits. The development of baits with these characteristics requires testing under conditions that maximize transfer of the material between segments of the cockroach populations.

Contact Effects With No Consumption. The effect of direct contact with bait without consumption can also provide better understand on how different baits can affect cockroach populations. These experiments are difficult because they require the sealing of mouthparts in cockroaches so they cannot consume the bait. These experiments produce better results when the insect life stage used is sufficiently resistant to lack of water, or is placed in an ambient environment were water loss does not cause serious mortality in the test population. Adult male German cockroaches have been used in such experiment due to superior survivability without feeding.

Once their mouthparts are sealed using a droplet of melted paraffin wax or other non-lethal method, the cockroaches are placed into arenas containing pre-weighed portion of a bait and other materials used for the direct consumption experiments. Mortality and other parameters can be observed during a short period of time (2-5 days) while the insects survive despite the lack of food and water consumption. These experiments have short durations due to the need to evaluate treatment mortalities within the time period when mortality in the control insects is still within reasonable levels. Beyond 3-5 days, mortality in the control insects will increase rapidly, mostly due to the lack of water in insects with their mouthparts sealed, and any results will be heavily influenced by that the control mortality.

Contact mortality with baits, besides being difficult to document, may be of lesser importance in population control, especially because cockroaches cannot survive long without ingesting food or water. Contact mortality relies on cockroaches investigating the bait without consuming it as would occur with bait-averse populations [19, 20]. Certain active ingredients, such as fipronil, are more likely to cause higher contact mortality than others; however, differences in bait matrices may also cause varying levels of contact mortality [19], and some formulations may have the potential for killing high proportions of cockroaches by contact alone.

Although different pest cockroach species have varying food preferences and, within the same species, there is great variability in the amount of bait individual cockroaches may consume, baits remain the most efficient method for control of cockroaches. Although maximization of bait consumption must take priority in bait product development, other factors that enhance secondary mortality and contact toxicity must be considered. Evaluation of these bait products in relation to direct mortality, by bait consumption, as well indirect mortality, by secondary and even tertiary contact with the active ingredient in the bait, are also important in development and evaluation of cockroach baits.

3. Ant bait evaluations

The control of social insects using baits relies greatly on the fact that in these social colonies, mortality of the individual workers has little effect on the survival of the colony. It is only by removing the reproductives, or at least a sufficient number of workers and juveniles to directly affect the reproductive potential of the colony, that an pest ant colony can be eliminated. Some of the active ingredients used in baits formulated for ant control completely bypass any effect on the worker, and concentrate their power with specific chemicals that interfere with the reproductive potential of the queen or queens. Because the reproductive individuals in ant colonies do not normally gather food or consume material that has not been somehow prepared by other colony individuals, reaching the reproductives is the greatest obstacle for any active ingredient formulated in an ant bait.

Because ant workers do not ingest large solid particles, ant bait formulations that target most urban ant pests must contain a liquid component. An ant head dissected shows the structures that prevent solids larger than 0.5 microns from being ingested (Figure 5). Food enters through the mouth and passes into the infrabuccal pocket. The infrabuccal pocket is a location for food particles too large to swallow. Food in the infrabuccal pocket passes through the buccal tube that is lined with setae that serve as filters. Particles too large to pass through this filtering mechanism remain in the infrabuccal pocket. These food particles can later be transferred to larvae for that ingest and digest these particles. Liquids that are ingested pass through the buccal tube into the pharynx and down the esophagus to the crop and midgut for storage and digestion. For baits to be ingested by urban pest ants they are usually liquids or granules that are soaked with liquid baits.

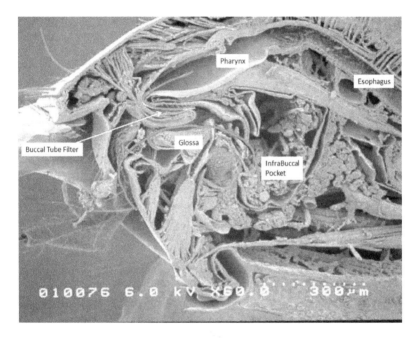

Figure 5. Cross section of an carpenter ant head showing the structures associated with ingestion of food.

Baits that target fungus-gardening ants target ants in a very different way, and are therefore develop following a completely different model. Most urban ant baits come in liquid, gel or granular formulations, but the granular formulation consists of a matrix containing a liquid that can the removed by the ants. Gel baits are only appropriate for indoor use or other special situations where protection from the climatic conditions is possible. Liquid baits normally require application into a holding device which the ants will have access to the bait. Granular baits are more practical for application outdoors, although they are also convenient for indoor applications.

Baits work by taking advantage of ant biology and behavior such as social grooming and trophallaxis. Once the bait is discovered, the foraging ants pick of the bait and transport it back to the colony. The brood, especially late instars, may be important in the digestion sol-id bait particles into a liquified form that can the transferred to workers and reproductives in the colony. The amount of brood in the colonies, in the laboratory and field, could be re-sponsible for the foraging preference. Fourth instar larvae do most of the protein digestion in the ant colony [21, 22] and their presence in a colony can change ant foraging preference to proteinaceous materials.

It is through food sharing that the toxicant in the bait can be transferred to the rest of the colony. Because the bait is picked up directly by the ant workers and is later shared within the colony, relatively low amounts of the toxicant can be used in targeting a pest ant popula-tion [23]. Ant foragers are usually the older workers that first pick up or consume the bait. They share the toxicant with other workers and queen tenders, and eventually after 3-4 days the toxicant reaches the queen, which affects reproduction in the colony. Even if the queen dies, eggs may hatch, larvae may pupate and develop into workers. Eventual control of a large colony may take 1-5 months.

The current baits out on the market for ant control include gel baits, liquid baits, and solid granular baits. A liquid or gel bait is usually one that requires a bait station and constant reapplication due to the elements and are usually used with ants that display mass recruit-ment to food sources.

In many cases there is little distinction between liquid and solid baits in terms of what the ants actually harvest in the field, as in the instance of popular fire ant baits. Fire ant baits consist of oil placed on a carrier. Foraging worker ants only feed from liquids so workers only remove oil off the bait granule. Thus the granule serves as a vehicle for the toxicant and attractant but it may not be carried into the nest at all. The active ingredient will enter the colony as a liquid.

Because granular baits can be broadcast over larger areas, this is the preferred formulation to reach most of the ant species. Granular baits take advantage of foraging patterns of differ-ent ant species. Granular baits consist of attractants, a carrier and active ingredients [24]. Four characteristics are important in a granular bait: 1) delayed toxicity, 2) easy transfer among individuals in the colony, 3) non-repellent active ingredient, and 4) attractive formu-lation for the target ant species [25, 23].

1. Delayed toxicity: In most ants, only a small percentage of the worker population active-ly forages outside of the nest. The use of active ingredient with delayed toxicity can guarantee maximum distribution of the bait within the colony before the ants start showing signs of toxicity. If there is enough delay, the active ingredient will be fed to larvae and reproductives before foraging and food sharing activities are shut down in the colony. This guarantees mortality of different castes within the nest, and the elimi-nation of immatures.

2. Easy transfer: This should be applied both to the bait itself, so that it can be handled by a maximum number of individuals within the colony, but especially to the active ingre-dient. With fast and easy active ingredient transfer, most of the colony can receive a le-

thal dose of the active ingredient before initial gatherers are affected and start showing toxicity effects.

3. Non-repellency: Because baits rely on the pick up and transfer by workers in the ant colony, non-repellent materials will be more easily masked within the bait formulation. Repellent active ingredients could also be used if they could be sufficiently masked by formulations components so not to prevent rejection of the bait by foragers.

4. Attractive formulation: The first step in the bait use process is attracting foragers so they are enticed to seek the bait and carry it back to the nest. Attractants added to the bait can overcome other deterrent characteristics of the bait or avoid any defense behavior that would normally prevent ants from returning toxic components to the nest. A great deal of work is dedicated to examining ant preferences to bait components so formulations can be picked up preferentially by the ants in an environment that will likely have many other food sources.

Several aspects of the granular bait formulation should be considered during the development of new products, including granule composition and size, attractive additives, and active ingredient.

The ideal granular bait should contain granules of similar size that can be easily applied to areas when needed. The carrier of the active ingredient is the most important part of the granular formulation because both the particle size and the materials and components used determine the spreading characteristics, the effectiveness of recruitment and removal of the bait, and the residual life of the active ingredient [26].

Size of the granular carrier may determine the size of ants that can be targeted with a particular formulation (Figure 6). In general, smaller ant species prefer smaller particle sizes to larger ones when given a choice [27]. If the particle size can be matched to a particular target pest ant species, this can increase the efficacy of the granular bait. Ants can normally carry granules with size that roughly match the size of head of foraging workers, but in some species, ant workers may collaborate in carrying larger pieces. Also, ants may subdivide larger bait particles before carrying the bait back to the nest. However, for baits developed for different ant species, this size matching may not be a preferred option.

A large granule size is more convenient because more active ingredient can be added to a larger particle size, allowing more active ingredient to be introduced into the colony with fewer particles collected by the foragers. Ant foraging normally fits what has been described as the optimal foraging theory [28, 29], which states that ants should take the biggest pieces of food particles that they can carry, in order to increase their net energy intake per unit of effort (Figure 7).

However, the difficulty in transport by the ants navigating the larger granular bait into the nest must be considered, especially for ant species that do not cooperate during foraging.

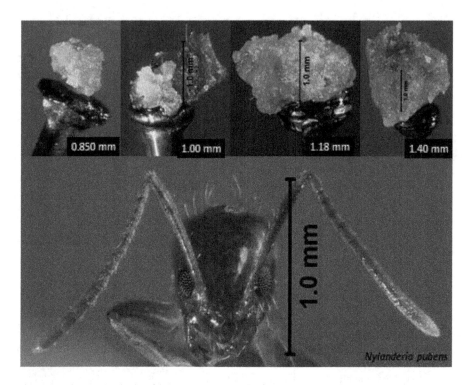

Figure 6. Workers head, *Nylanderia fulva*, in comparison with dog food granules used in size preference experiments.

Because different ants have different requirements for protein, lipids and sugar, and these requirements are likely to vary throughout the year or the life of the ant colony, the composition of the granule can be critical in bait development. Two approaches have been used in the formulation of granular baits for ants: a) use of non-nutritious granule to which food attractants are added, b) use of food particles as the granule matrix to which the active ingredient is added. With either approach, the quality of the food attractant determines which ants are attracted to the bait. Although several bait compositions use sugars as the only attractant, some products have been formulated with protein and lipid attractants, and even insect tissue.

A non-nutritious granule normally used in formulating ant baits is de-fatted corncob, a byproduct from corn processing [30, 24]. It is capable to absorbing relatively large amounts of the liquid additives such as the oil used in many fire ant baits. Fine granules from dog food or other animal diets serve as a nutritious granule for ant baits because it fulfills ant nutrient requirements, is easy to prepare in a uniform granular size and it readily absorbs additives.

Figure 7. *Nylanderia fulva* foraging in a laboratory setting on 1.00-mm (Top) and 1.40-mm (Bottom) dog food granules used in size preference experiments.

The addition of insect tissue [31], which attempts to mimic the natural diet of many ant species, adds complexity to the bait but also to the production process, with inevitable cost consequences. The use of a readily available byproduct, such as silkworm pupae can facilitate production and cut costs. Other insects that can be mass reared at low costs, such as crickets or waxmoth larvae, are also interesting alternatives. Laboratory reared crickets ground into a slurry with addition of small quantity of water have been used in our laboratory to be added to dog food granules tested as a bait to tawny crazy ant (*Nylanderia fulva*).

Traditional baits for *S. invicta*, consist of oil (attractant) on corn cob matrix (carrier) [30, 24]. On the other hand, baits that contain proteins and carbohydrates are very attractive to species such as *L. humile* and *Paratrechina* spp. that are not attracted to the lipid-based fire ant baits [24].

In order to enhance a carrier, it is important to know the ant species food preferences, based on field observations, laboratory experiments, and reports in the literature on similar ant species. Observations of feeding habits such as ants feeding on honeydew from aphids, plant nectaries and insect tissue [32], tending to aphids and mealy bugs [33] can serve as clues in the development of ant baits. Preference studies indicating the balance between components in the ant diets and other aspects of ant nutrition [21, 34, 35, 36, 37, 38, 39] also offer valuable clues that can help the development of new ant baits.

Ant Bait development. Bait development should be initiated with preference tests in the laboratory (Figure 8), but should quickly move into field tests due to the great variability food preference and gathering between laboratory and field populations of the same ant. Laboratory colonies are usually fed a constant diet that is rich in all nutrients need by the ants, while field ant populations are more likely to go through periods when their diet is relatively low in certain nutrients or components such as live insects or sugars. Differences between controlled environment in the laboratory and more variable environment in the field lead to differences in foraging behaviors preference to bait components [40, 41, 42, 43]. Differences in the presence and proportion of different developmental life stages in the colony [44] can also be important factors in determining differences between laboratory and field results.

Field tests should also be conducted at different times of the year in order to characterize the bait preference and effect given different levels of foraging on a specific formulation throughout the season. Because foragers need to move very little between the nest and the foraging arena in laboratory colonies, food choices may be different from those for field ants which usually will travel much further from the nest both in scouting for new food sources and in foraging trails. Distance between the nest and food source can affect choice and quantify of food gathered by an ant colony.

In the laboratory, arena preference tests with multiple bait choices can be used in the elimination of candidate formulations. Later tests with limited choices can be used later to further define preference for specific formulations. Careful design of the foraging arena will avoid preference biases for baits that are found more readily.

If using colony fragments in foraging experiments, careful attention to the size and composition of these experimental colonies is important to preserve the foraging behavior and other characteristics that match those of the full colonies (Figure 9). For instance, foraging of different baits can be drastically affected by the presence or absence of brood, and the brood age structure in an ant colony.

Figure 8. Testing arena used for experiments on *Nylanderia fulva* using granular bait matrix applied with active ingredient.

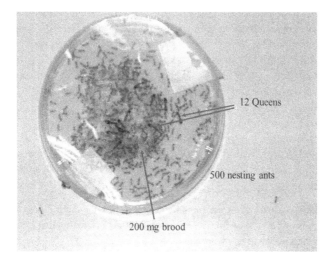

Figure 9. Colony fragments of Pharaoh ants are set up in cells containing brood, workers and queens.

Eventually, bait development must include field experiments to determine the fate and efficacy of the material when applied in situations for which the bait is designed. Careful observations on ant behaviors associated with finding, gathering, and moving the bait material to the ant colony are useful in understanding potential shortcomings of the developed products. Use of fields in different locations that represent the variety of situations where the ant baits can be used is essential in defining clearly the effects of different parameters and factors on the performance on the ant bait.

The data collected from the different experiments will vary, and should be adjusted for each ant species and situation. However, at minimum, the data should allow estimation of the quantity of material necessary to eliminate pest ant colonies, and the total mortality after different periods of time. Of particular importance, is the effect that bait may have on reproductive individuals. In polygyne colonies, it is especially important that the bait achieves maximum distribution within the colony, so that reproductive individuals throughout the colony can be effectively controlled.

4. Termite bait evaluations

Methods of subterranean termite exclusion and prevention of structural infestations have broadened from soil termiticides and barrier treatments to include monitoring and baiting systems. Baiting systems have increased in registration and use since the introduction of the first bait 18 years ago (tradename Recruit, Dow AgroSciences LLC, Indianopolis). The specificity and mode of action of these active ingredients requires much less product to be applied to the environment. Hexaflumeron was the first active ingredient (AI) registered in the United States to be used in a termite bait formulation and there are currently several other AI in use, all of which fall into two classes: insect growth regulators (IGR) and energy production inhibitors. Both classes are considered to be slow-acting and rely on foraging termites to transfer small amounts of consumed bait material thoughout the colony though contact, trophallaxis, grooming, fecal consumption and cannibalism. Baiting systems using IGRs are intended to be used as stand alone treatments. Bait formulations with AI affecting energy production are used in conjunction with soil treatments. There are baits designed to be used in-ground to prevent structural infestation and others for use above-ground in areas with known termite activity. A successful baiting system should be proven to affect termite populations (Figure 10).

Active ingredient evaluation. A non-repellent, lethal and slow-acting active ingredient is required for a termite bait to be effective. When evaluating a potential bait toxicant one must first determine the toxicity of it towards the termite species it will be used againt. For example, at what concentration will it kill 90% of exposed termites (LC_{90})? How long does it take for that 90% to die (LCt_{90})? Termites are highly social and do not fare well when kept in small numbers so great numbers of exposed termites should be placed together

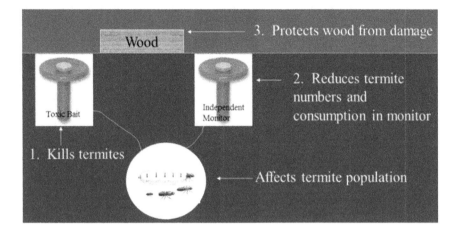

Figure 10. Termite bait efficacy data can document termite populations are affected in 3 ways.

Once the ideal bait toxicant concentration has been determined it must be tested to determine whether it will be readily consumed or show a feeding deterrent effect when adhered to wood or other cellulose containing matrix.

Bait formulation evaluation. The formulation of a palatable bait matrix is essential to the success of a baiting program. In the natural environment of subterranean temites there are a lot of potential food sources providing competition for baits. If there is no preference shown towards a bait it will not be very successful when installed below ground. Above ground baits must be more palatable because the termites present already have a source of wood that they are consuming. Impregnating wood or another cellulose containing material with active ingredient is the most common formulation method for commercially available baits.

Choice tests can be used to determine termite feeding preferences and any foraging biases. In a laboratory setting one can prepare a trial arena filled with moistened soil (10% wt:wt). Each dish will contain one piece of untreated wood and one piece of bait matrix, both of which have the same moisture content, dimensions and orientation within the arena. One thousand subterranean termites are introduced in a Petri dish with a small opening covered by a piece of filter paper at the end. Once the filter paper has been consumed the termites will be free to tunnel and forage throughout the arena. Locations of food choices should be randomized to eliminate any directional biases. Repeating this experiment multiple times with multiple colonies will show which food is preferred by termites. Which source did the termites consume the most of by weight? Was the first food source contacted the only one consumed or was there cross over between both food sources? Statistical analysis will indicate if termites prefer the bait matrix to untreated wood. This simple assay can be repeated and altered to include multiple wood and bait choices, different soil types and moisture levels.

Baiting system evaluation. Using a baiting program to prevent structural infestation may take considerably longer to be effective than soil treatments and thus the methods used to evaluate them are different. In order to gain registration, evaluations of termite baits must ultimately fulfill requirements set forth by state government guidelines (*Florida: 5E-2.0311 Performance Standards and Acceptable Test Conditions for Preventive Termite Treatments for New Construction*). Stand-alone baiting systems must be tested and meet specific requirements in field plot and building tests.

Independent Monitors	Building Monitoring	Reinfestation of Buildings
• >90% reduction in termite activity	• Cessation of live termite activity	• Visual inspection showing no reinfestation within 2 years
• >90% of test biuldings protected	• >90% of test biuldings protected	• Research and visual inspection showing no reinfestation within 1 year
• Protection within 12 months of initiation of feeding on bait active ingredient	• Protection within 12 months of initiation of feeding on the formulated bait	

Table 1. Performance standards for stand-alone termite baits in structures with existing infestations.

Evaluation of below-ground baits: field plot tests. Once it has been determined that feeding on bait has started, infested field plot tests require a reduction in each termite population by at least 50% or a reduction of wood consumption at independent monitors by a minimum of 50% in at least 75% of baited population colonies within one year.

The minimum required thresholds must be maintained for at least 6 months. In order to meet this requirement one can place monitoring stations, which have the same shape, appearance and moisture levels of a baiting station but contain untreated wood instead of bait, throughout a field plot or around a structure known to contain termites. Monitoring stations containing untreated wood are installed in the ground in augered holes at consistent intervals (every 10-20 feet).

Monthly inspections of monitoring stations will continue until live termites are found. Monitoring stations without termites are not switched to bait stations until live termites are found. When live termites are found monitoring stations are deemed 'active', wood will be replaced by a bait tube containing active ingredient and termites contained in the wood will be placed in the bait station. Plastic bucket traps (with uniform holes allowing for termite entry) containing wooden blocks may be placed around these stations and checked monthly. The purpose of these bucket traps is to be able to count and assess the nature of termite activity and chart differences over time. Commercially available bait stations require different monitoring intervals but for evaluation it is recommended to be conducted monthly to better determine when control has been achieved.

During monthly inspections, the number of termites present and the amount of bait consumed will be recorded. Bait matrix consumption is typically a visual estimate of the percent consumed as bait weights can be misleading. If baits are completely consumed, compromised or damaged they will be replaced with new bait. Once termite presence and bait consumption ceases monitoring resumes and monthly inspections will continue for at least 6 months. If monitoring stations are found to be active a new bait tube till be installed. The amount of active ingredient consumed can be measured at the end of the study by drying the bait and comparing initial and post-treatment weights using the percentage of active ingredient by weight in the matrix formulation.

The question of whether a colony has been eliminated of merely suppressed can be difficult to answer and may require months to years of monitoring before and after a baiting system is put in place. A suppressed colony will exhibit a period of inactivity in which no termites will be found in monitoring stations yet eventually recover and continue foraging [45]. To better determine the level of control achieved it is recommended to use cuticular dyes and genetic markers as detailed below.

Detecting presence of multiple colonies and foraging areas. Once a monitoring station is attacked by termites they will be collected and keyed to species level. If one is interested in determining the number of colonies present and their respective foraging areas in a field site there are two options: mark-release-recapture or cuticular dyes. Mark-release-recapture involves collecting live termites at monitoring stations, bringing them into the laboratory and feeding them filter paper impregnated with fat-soluble cuticular dyes such as Nile Blue A, Sudan Red 7B, or Neutral Red (Fisher,Pittsburgh, PA) [46]. Once termites are dyed they are placed back into the stations from which they are collected. Mark-release-recapture will also help in estimating population sizes.

A less obtrusive method involves placing a cellulose matrix impregnated with cuticular dyes, Nile Blue A or Neutral Red, in bait tubes which will allow for long-term tracking of termites from station to station. This eliminates the need for termites to be handled. Once the termites are dyed and back in test site the procedure is the same. During monthly monitoring of stations the locations and numbers of dyed termites can be recorded and a map of foraging activities produced [47]. Both methods can be enhanced though the use of genetic markers to help differentiate between colonies [48, 49, 50].

Evaluation of above-ground baits: building tests. Evaluation of above-ground baits can only be conducted in buildings with active subterranean termite infestations. Mud tubes are broken and both monitoring and bait stations are installed in line with the disturbed tube. Weekly monitoring is recommended because bait toxicants are introduced to the colony very quickly. Bait consumption will be visually estimated as a percentage and should be replaced if too much has been consumed or the bait has been compromised. If baits are too dry water may be added but too much water will become a deterrent to consumption. Termites will be counted in monitoring stations but not removed. Once feeding on bait and monitoring stations has ceased, baits will be replaced with monitoring stations.

Above-ground baiting programs must show ≥ 90% reduction in termite activity in ≥ 90% of test buildings within one year from the initiation of bait consumption. A successful above-ground baiting program must show that there has been no re-infestation within one year after activity has ceased. This must be verified by combining a visual inspection with termite detecting tools including infrared devices, moisture meters, radar, chemical detection, bath trap inspection ports, canine detection or fiber optics. The alternative is to wait until two years after the last evidence of termite presence and conduct a visual inspection of the site.

5. Summary

Baits have many advantages for use in urban environments. The advantages extend from use in IPM programs, to non-impact of humans who are living and living or working amid infestations, and to advantages associated with controlling the pests (Table 2).

IPM	Human	Pest Control
• Preserve beneficial organisms involved in biological control	• No odor	• Slow acting
• Reduced risk	• Lower exposure to pesticides	• Non-repellent
• Can be used in sensitive areas	• No mixing needed	• Attractives and phagostimulants to enhance consumption
• Long lasting	• Less preparation prior to pesticide application	• Secondary mortality by transfer
• Application as point sources		• Long lasting
• Less active ingredient		• Transfer of AI within pest population
• Narrow spectrum of insects controlled		• Translocation • Overcome insecticide resistance

Table 2. Advantages of baits in pest management in urban environments in relations to IPM principles, humans and pest control.

Insecticides used in urban environments are almost always in proximity to people, pets, and food. As a result, the safety of products used and efficacy of the formulation in urban pest control are of extreme importance. People can be affected by the use of a wrong formulation, or buildings can be destroyed when ineffective products are used. As a result of screening active ingredients and formulation, a variety of insecticides have been developed for urban pest management (Table 3). Most of the active ingredients are listed by the USEPA as reduced risk products. As reduced risk, there is an expedited registration process for baits containing these actives.

Type/Active Ingredient	Mode of Action	Pest Groups
• Oxadiazine Indoxacarb	• Sodium channel blockage	• Cockroaches, ants
• Neonicotinoid Imidacloprid Dinotefuran	• Acetylcholine receptor stimulation	• Cockroaches, ants, flies
• Spinosins Spinosad	• Acetylcholine receptor stimulation	• Flies
• Phenylpyrazoles Fipronil	• GABA receptor blockage	• Cockroaches, ants
• Avermectins Abamectin Emamectin	• Glutamate receptor stimulation	• Cockroaches, ants
• Chitin synthesis inhibitors Hexaflumuron Noviflumuron Diflubenzuron	• Block chitin formation	• Termites
• Amidinohydrazone Hydramethylnon	• Inhibit energy production	• Cockroaches, ants
• Pyrroles Chlorfenapyr	• Inhibit energy production	• Termites
• Borates Boric acid Sodium borate Disodium octaborate tetrahydrate	• Non-specific metabolic disruption	• Cockroaches, ants, flies

Table 3. Active ingredients, modes of action, and pests controlled with baits used for pest management in urban environments.

Baits have become one of the most popular formulations used by pest management professionals for use against cockroaches, ants, and termites. One of the advantages of bait formulations is that they are usually ready to use, in low concentrations, and can be placed only where and when needed. Hazards of using baits is minimized by using child-resistant bait stations or careful placement directly into harborages. The use of baits requires more time than spraying and costs may be higher because of the use of food-grade ingredients in the formulation.

Overall, baits are a very effective and successful insecticide formulation for urban pest control. As a result, the industry has been expanding testing and screening programs for label expansions so insects other than ants, cockroaches, and termites can be controlled.

Author details

Bennett W. Jordan, Barbara E. Bayer, Philip G. Koehler and Roberto M. Pereira

Department of Entomology and Nematology, Steinmetz Hall, Natural Area Drive, University of Florida, Gainesville, FL, USA

References

[1] Bell WJ, Roth LM, Nalepa CA. Cockroaches: ecology, behavior, and natural history. Baltimore: Johns Hopkins; 2007.

[2] Koehler PG, Atkinson TH, Patterson RS. Toxicity of abamectin to cockroaches (Dictyoptera: Blattellidae, Blattidae). Journal Economic Entomology 1991;84: 1758-1762.

[3] Silverman J, Bieman DN. Glucose aversion in the German cockroach, *Blatella germanica*. Journal of Insect Behavior 1993;11: 93-102.

[4] Wang C, Scharf ME, Bennett GW. Behavioral and physiological resistance of the German cockroach to gel baits (Blattodea: Blattellidae). Journal of Economic Entomology 2004;97: 2067-2072.

[5] Valles SM, Strong CA, Koehler PG. Inter- and intra-instar food consumption in the German cockroach, *Blattella germanica*. Entomologia Experimentalis and Applicata 1996;79: 171-178.

[6] Koehler PG, Strong CA, Patterson RS, Valles SM. Differential susceptibility of German cockroach (Dictyoptera: Blattellidae) sexes and nymphal age classes to insecticides. Journal of Economic Entomology 1993;86: 785-792.

[7] Appel AG. Performance of gel and paste bait products for German cockroach (Dictyoptera: Blattellidae) control: Laboratory and field studies. Journal of Economic Entomology 1992;85: 1176-1183.

[8] Appel AG. Laboratory and field performance of an indoxacarb bait against German cockroaches (Dictyoptera: Blattellidae). Journal of Economic Entomology 2003;96: 863-870.

[9] Durier V, Rivault C. Secondary transmission of toxic baits in German cockroach. Journal of Economic Entomology 2000;93: 434-440.

[10] Gahlhoff Jr JE, Miller DM, Koehler PG. Secondary kill of adult male German cockroaches (Dictyoptera: Blattellidae) via cannibalism of nymphal fed toxic baits. Journal of Economic Entomology 1999;92: 1133-1137.

[11] Kopanic Jr RJ, Schal C. Coprophagy facilitates horizontal transmission of bait among cockroaches (Dictyoptera: Blattellidae). Environmental Entomolology 1999;28: 431-438.

[12] Buczkowski G, Schal C. Emetophagy: fipronil-induced regurgitation of bait and its dissemination from German cockroach adults to nymphs. Pesticide Biochemistry and Physiology 2001;71: 147-155.

[13] Buczkowski G, Schal C. Method of insecticide delivery affects horizontal transfer of fipronil in the German cockroach (Dictyoptera: Blattellidae). Journal of Economic Entomology 2001;94: 680-685.

[14] Buczkowski G, Scherer CW, Bennett GW. Horizontal transfer of bait in the German cockroach: indoxacarb causes secondary and tertiary mortality. Journal of Economic Entomology 2008;101: 894-901.

[15] Holbrook GL, Roebuck J, Moore CB, Waldvogel MG, Schal C. Origin and extent of resistance to fipronil in the German cockroach, *Blattella germanica* (L.) (Dictyoptera: Blattellidae). Journal of Economic Entomology 2003;96: 1548-58.

[16] Appel AG, Reierson DA, Rust MK. Comparative water relations and temperature sensitivity of cockroaches. Comparative Biochemistry and Physiology Part A: Physiology 1983;74: 357-361.

[17] Gunn DL. The temperature and humidity relations of the cockroach. III. A comparison of temperature preference, and rates of desiccation and respiration of *Periplaneta americana, Blatta orientalis* and *Blatella germanica*. Journal of Experimental Biology 1935;12: 185-190.

[18] Waldbauer GP, Friedman S. Self-selection of optimal diets by insects. Annual Review of Entomology 1991;36: 43-63.

[19] Buczkowski G, Kopanic Jr. RJ, Schal C. Transfer of ingested insecticides among cockroaches: effects of active ingredient, bait formulation, and assay procedures. Journal of Economic Entomology 2001;94: 1229-1236.

[20] Metzger R. Behavior. In: Rust MK, Owens JM, Reierson DA (eds.) Understanding and Controlling the German Cockroach. New York: Oxford University Press. 1995. p49-76.

[21] Petralia RS, Sorensen AA, Vinson SB. The labial gland system of larvae of the imported fire ant, *Solenopsis invicta* Buren. Cell Tissue Research 1980;206: 145-156.

[22] Weeks Jr RD, Wilson LT, Vinson SB, James WD. Flow of carbohydrates, lipids, and protein among colonies of polygyne red imported fire ant, *Solenopsis invicta* (Hymenoptera: Formicidae). Annals of the Entomological Society of America 2004;97: 105-110.

[23] Hooper-Bùi LM, Rust MK. Oral toxicity of abamectin, boric acid, fipronil, and hydramethylnon to laboratory colonies of Argentine ants (Hymenoptera: Formicidae). Journal of Economic Entomology 2000;93: 858-864.

[24] Stanley MC. Review of the efficacy of baits used for ant control and eradication (2004 Landcare Research Contract Report: LC0405/044). http://argentineants.landcarere-

search.co.nz/documents/Stanley_2004_Bait_Efficacy_Report.pdf. (accessed 9 Sep 2012).

[25] Stringer Jr CE, Logren CS, Bartlett FJ. Imported fire ant toxic bait studies: evaluation of toxicants. Journal of Economic Entomology 1964;57: 941-945.

[26] NPCA. Granules. National Pest Control Association Technical release 1965;5-65:1-5.

[27] Hooper-Bùi LM, Appel AG, Rust MK. Preference of food particle size among several urban ant species. Journal of Economic Entomology 2002;95: 1222-1228.

[28] Nonacs P, Dill LM. Mortality risk vs. food quality trade-offs in common currency: ant patch preferences. Ecology. 1990;71: 1886-1892.

[29] Roulston, TAH, Silverman J. The effect of food size and dispersion pattern on retrieval rate by the Argentine ant, *Linepithema humile* (Hymenoptera: Formicidae). Journal of Insect Behavior 2002;15: 633-648

[30] Lofgren CS, Bartlett FJ, Stringer CE. Imported fire ant toxic bait studies: evaluation of carriers for oil baits. Journal of Economic Entomology 1963;56: 63-66.

[31] Williams DF, Lofgren CS, Vander Meer RK. Fly pupae as attractant carriers for toxic baits for red imported fire ants (Hymenoptera: Formicidae). Journal of Economic Entomology 1990;83: 67-73.

[32] Creighton WS. The ants of North America. Bulletin of the Museum of Comparative Zoology of Harvard College. 1950;104: 1-585.

[33] Wetterer JK, Keularts JLW. Population explosion of the hairy crazy ant, *Paratrechina pubens* (Hymenoptera: Formicidae), on St. Croix, US Virgin Islands. Florida Entomologist 2008;91: 423-427.

[34] Cook SC, Wynalda RA, Gold RE. Macronutrient regulation in Rasberry crazy ant (*Nylanderia* sp. nr. *pubens*). Insectes Sociaux. 2012;59: 93-100.

[35] Dussutour A, Simpson SJ. Communal nutrition in ants. Current Biology 2009;19: 740-744.

[36] Vogt JT, Grantham RA, Corbett E, Rice SA, Wright RE. Dietary habits of *Solenopsis invicta* (Hymenoptera: Formicidae) four Oklahoma habitats. Environmental Entomology 2002;31: 47-53.

[37] Harris R, Abbott K, Berry J. Invasive ant threat: *Anoplolepis gracilipes*. Information sheet. Landcare Research, Manaaki Whenva. http://www.landcareresearch.co.nz/research/biocons/invertebrates/Ants/invasive_ants/anogra_info.asp. (accessed 12 July 2012).

[38] Kenne M, Mony R, Tindo M, Njaleu LCK, Orivel J, Dejean A. The predatory behavior of a tramp species in its native range. Comptes Rendus Biologies 2005;328: 1025-1030.

[39] Pagad S. Issg database: ecology of *Paratrechina longicornis*. National Biological Information Infrastructure & Invasive Species Specialist. http://www.issg.org/database/species/ecology.asp?si=958. (accessed 12 July 2012).

[40] Traniello, JFA. Social organization and foraging success in *Lasius neoniger* (Hymenoptera: Formicidae): Behavioral and ecological aspects of recruitment communication. Oecologia 1983;59: 94-100.

[41] Vogt JT, Smith WA, Grantham RA, Wright RE. Effects of temperature and season on foraging activity of red imported fire ants (Hymenoptera: Formicidae). Environmental Entomology 2003;32: 447-451.

[42] Challet M, Jost C, Grimal A, Lluc J, Theraulaz G. How temperature influences displacement and corpse aggregation behaviors in the ant *Messor sancta*. Insectes Sociaux 2005;52: 309-315.

[43] Wiltz BA, Suiter DR, Berisford W. A Novel delivery method for ant (Hymenoptera: Formicidae) toxicants. Midsouth Entomologist 2010;3: 79-88.

[44] Traniello JFA. Foraging strategies of ants. Annual Review of Entomoly 1989;34: 191-210.

[45] Su, N-Y, Scheffrahn RH. A review of subterranean termite control practices and prospects for integrated pest management programmes. Integrated Pest Management Reviews 1998;3: 1-13.

[46] Messenger, MT, Su N-Y, Husseneder C, Grace JK. Elimination and reinvasion studies with *Coptotermis formosanus* (Isoptera: Rhinotermitidae) in Louisiana. Journal of Economic Entomology 2005;98: 916-929.

[47] Atkinson, TH. Use of dyed matrix in bait stations for determining foraging territories of subterranean termites (Isoptera: Rhinotermitidae: *Reticulitermes* spp., and Termitidae: *Amitermes wheeleri*). Sociobiology 2000;36: 149-167.

[48] Vargo, EL. Genetic structure of *Reticulitermes flavipes* and *R. virginicus* (Isoptera: Rhinotermitidae) colonies in an urban habitat and tracking of colonies following treatment with hexafumuron bait. Environmental Entomology 2003;32: 1271-1282.

[49] Vargo, EL. Hierarchical analysis of colony and population genetic structure of the eastern subterranean termite, *Reticulitermes flavipes*, using two classes of molecular markers. Evolution 2003;57: 2805-2818.

[50] Vargo, EL, Henderson G. Identification of polymorphic microsatellite loci in the Formosan subterranean termite *Coptotermes formosanus* Shiraki. Molecular Ecology 2000;9: 1935-1938.

Chapter 7

The Use of Deltamethrin on Farm Animals

Papadopoulos Elias

Additional information is available at the end of the chapter

1. Introduction

Ectoparasites reduce significantly animal production and welfare. They cause nuisance, anaemia, irritation and transfer of pathogens of important diseases, often leading to animal death. Examples of diseases with high mortality transmitted by arthropods include viral diseases, such as the Bluetongue disease, or parasitic diseases, such as piroplasmosis and filariosis. Biting midges of the *Obsoletus* species complex of the ceratopogonid genus *Culicoides* were assumed to be the major vectors of bluetongue virus in northern and central Europe during the 2006 outbreak of bluetongue disease. Most recently, field specimens of the same group of species have also been shown to be infected with the newly emerged Schmallenberg virus in Europe, as other bloodsucking arthropods. Furthermore, ectoparasites may attack humans and threaten public health, such as diseases transmitted by mosquitoes or ticks.

The control of ectoparasites found on animals, i.e. midges, fleas, ticks, lice, flies, is largely based on the use of chemicals (insecticides). The main groups, which have been used as the basis of the common ectoparasiticides, include the synthetic pyrethroids, organochlorines, organophosphates, carbamates, formamidines and others. The macrocyclic lactones (avermectins and milbemycines) have also been shown to have a high activity against a range of ectoparasites. Furthermore, there are also compounds which affect the growth and development of insects, such as the chitin inhibitors, chitin synthesis inhibitors and juvenile hormone analogues. Insect growth regulators (i.e. lufenuron) are used mostly against fleas and certain flies.

2. Deltamethrin

Pyrethroids, synthetic analogues of pyrethrins, were developed to improve stability of the natural pyrethrins since they degraded rapidly by light. The pyrethrin insecticides were originally derived from extracts of the flower heads of *Chrysanthemum cinerariaefolium*. There are six compounds that comprise the natural pyrethrins, namely, pyrethrins I and II, cinerins I and II, and jasmolins I and II.

Pyrethroid insecticides are attractive compounds because of their high potency and ability to reduce disease transmission, selective toxicity, relative stability in the environment and ease of degradation in vertebrates. Compared with organophosphates and carbamate insecticides, pyrethroids are less likely to cause acute and chronic health effects to vertebrates. The common synthetic pyrethroids in use include deltamethrin, permethrin, cypermethrin, flumethrin and others. The main value of these compounds is their repellent effect and since they persist well on the coat or skin, but not in tissue, they are of particular value against parasites that feed on the skin surface such as ticks, lice, some mites and nuisance flies. They can act as contact insecticides due to their property to be lipophilic. Some have the ability to repel and to affect flight and balance without causing complete paralysis (knockdown effect). They pose a strong affinity for sebum. They are widely used in veterinary medicine for agricultural and domestic purposes.

Pyrethroids are primarily targeted on the nervous system. They act as neurotoxins upon sensory and motor nerves of the neuroendocrine and CNS of arthropods. Several mechanisms of action have been proposed, including alterations in sodium channel dynamics in nerve tissues, which polarise membranes and result in abnormal discharge in targeted neurons.

Synthetic pyrethroids are relatively safe. However, if toxicity occurs, it is expressed in the peripheral nervous system of animals as hypersensitivity and muscle tremors. They are extremely toxic to fish and aquatic invertebrates (except for molluscs and amphibians). However, it would appear that, in practice, the risks of deltamethrin are limited. Light, the pH of the water, organic or colloidal molecules in suspension, and the presence of sediment and bacteria, all contribute to a rapid breakdown of the molecule into rapidly decomposed non-toxic products. Deltamethrin does not present any toxicity problems for birds, including game birds. Regarding public health, some adverse effects on humans may occur, with neurotoxicity and developmental toxicity being potential side effects following acute high-dose exposures to pyrethroids.

Among synthetic pyrethroids, Deltamethrin (Butox, MSD) is of particular importance. Contrary to other pyrethroids, it is a single *cis*-isomer (Figure 1), which is considered to be more effective than isomer combinations. Deltamethrin repels ectoparasites by the "hot foot effect", which is typical for pyrethroids. An insect after it had a "touchdown" on such an animal, redraws its feet suddenly from treated hair. Even after a very short contact, for only a few seconds, to treated hair, a "knock-down effect" occurs since insects and ticks die soon after the open nerve ends at their feet got into contact with the insecticide. This efficacy leads to a constant reduction of biting or attacks and same time dead female population stops breeding. On the other hand, studies carried out in 3 generations of rats, using daily doses of 0.15, 1 and 3.75 mg/kg in the feed, did not reveal any differences between treated and control animals with respect to fertility, duration of gestation, fecundity and viability of the litters. Finally, Deltamethrin did not show any mutagenic effects in any of the tests (both *in vivo* and *in vitro*) employed.

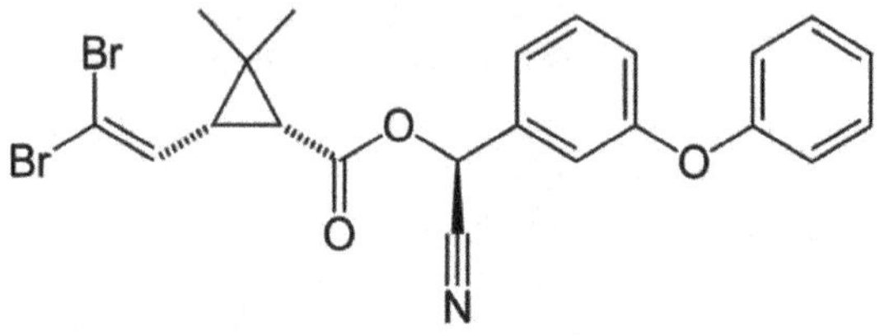

Figure 1. The structure of deltamethrin

3. Efficacy

Several published papers exist in scientific journals demonstrating the strong properties of Deltamethrin (Butox, MSD) to repel or kill arthropods infesting livestock, such as biting midges, nuisance flies, ticks, lice, certain mites etc. Results from these field trials proved the high efficacy of this compound to protect ruminants from midges, i.e. *Culicoides* spp., for periods over 4-5 weeks, even if the animals became wet several times. It has been found to be effective against ticks, including all developmental stages, mosquitoes and many others.

Herein follow, in more details, the results of some studies evaluating the potential use of Deltamethrin (Butox, MSD) on farm animals.

3.1. *Culicoides* midges

One set of such experiments include the investigation of the control of Bluetongue disease of ruminants using this drug, carried out by Schmahl and colleagues in 2008. Bluetongue disease is a viral disease, which harms considerably farm ruminants with high mortality rates in cattle and especially in sheep, while wild ruminants become infected, serve as virus reservoirs, but show only rarely severe symptoms of disease. From several transmission experiments and epidemiological studies in South Africa and in Southern Europe, it was known that the main vector belonged to the midges (Family Ceratopogonidae, genus *Culicoides*). Therefore, protection methods were needed to avoid transmission of the virus from one animal to the other. Thus, the aim of this study was to compare the efficacy of Deltamethrin when *Culicoides* specimens come into contact with hair of cattle and sheep that had been treated for 7, 14, 21, 28, or 35 days before. This study was needed, since it had to be clarified, whether the product in this formulation of Deltamethrin can reach the hair of feet in sufficient amounts when they are applied onto the hair along the back line. The product must arrive in sufficient amounts at the feet and along the belly since there are the predominant biting sites of the very tiny (only 0.8–3 mm long) specimens of *C. obsoletus*, *C. pulicaris* and *C. dewulfi*, the proven vectors of Bluetongue in Europe. Towards this end, one group of three young cattle of about 400 Kg bodyweight and one group of three young sheep of 60 Kg each were treated by application (pour on) of 30 ml and 10 ml, respectively, of the product Butox® 7.5 onto the skin along the backside of the animals. Butox® 7.5 contains 7.5 g deltamethrin per liter of the ready-to-use solution and is a registered trademark of MSD pharmaceutical company. Seven, 14, 21, 28, and 35 days after treatment, hair was clipped off from the feet of the cattle and sheep (just above the claws), collected in separate, suitable plastic bags, and transported to the institute, where it was mixed with freshly caught midges, which had been caught in the previous night with the aid of an ultraviolet light lamp. Each vial contained at least ten *Culicoides* specimens, besides other insects. The trapped insects were incubated with treated hair or with hair of an untreated animal (control). The exposure periods of the insects to hair lasted for 15, 30, 60, or 120 seconds—a period which was thought to be realistic compared to the field conditions. The insects were thereafter separated from the hair and placed on filter paper inside closed plastic petri dishes, where they were observed at regular intervals (5–10 min) using a stereo microscope to record reactions and the time of death after the first contact with treated hair.

The midges (*Culicoides* species) were apparently highly sensitive to Deltamethrin (Butox® 7.5) since they died even after rather short contacts to hair treated even 35 days ago, which is a very satisfying effect. There were no significant differences between the species of treated animals (sheep or cattle), although the distance from the place of application (back) until the feet is longer in cattle than in sheep. Thus, the formulation can reach in sufficient amounts the region of the predominant biting sites of the *Culicoides* species (feet, belly).

The results obtained from these experiments clearly show that Deltamethrin, when applied as a pour-on solution onto the back of the animals, has a significant killing effect on the *Culicoides* species, which are known vectors of the Bluetongue virus in Europe. Furthermore, even if the protection might not be 100%, any killed female *Culicoides* prevents its possible progeny and hinders the transmission of agents of diseases.

During the above described studies, the animals were provided with adequate shelter against the rain. The efficacy of deltamethrin on wet animals was, therefore, not tested. It is common that ruminants stay under the rain or exposed to water when they stay at pasture for grazing. Therefore, the same group of researchers in 2009, carried out the next step of the above study, i.e. the new task was to determine if thoroughly wetting the test subjects, twice a week, would affect the efficacy of Deltamethrin. Cattle and sheep were treated with Butox 7.5 along the neck or dorsal midline, as described earlier. Test animals were wet thoroughly with tap water twice weekly. Control animals remained dry. Hair was clipped off the legs, near the claws, at day7, 14, 21, and 28 after treatment of test and control animals. Recently caught *C. obsoletus* midges were then exposed to the hair for 15, 30, 60, and 120 seconds. The midges were then transferred to filter paper in plastic petri dishes and observed. The time needed for the midges to die after the exposure was recorded.

In both cattle and sheep, the product remained active for at least up to 4 weeks (28 days - end of the experiment), even in the animals wet with water twice weekly over the 4-week period. In sheep, the time between exposure and death of the midges was definitely lengthened in animals that were wet. In cattle, the results were different in that in some cases, time between exposure and death of the midges was shorter in wet animals than in dry animals, while in other cases the results were similar. Compared with the sheep, the time between exposure and death is generally quicker, probably, due to differences in hair structure. All midges exposed to hair from treated sheep or cattle, wet or dry, died even after only 15 seconds of exposure to the hair. In cases where the period between exposure and death of the midges was very long, it is likely that the midges were unable to bite as they showed signs of paralysis immediately after contact with treated hair. The fact that Butox 7.5 pour on remains effective in animals regularly exposed to rain is an important finding towards the protection of ruminants from the attack of midges and the risk of disease transmission.

3.2. Ticks

Another study of Mehlhorn and others took place recently in 2011 in order to investigate the efficacy of Deltamethrin (Butox® 7.5 pour on) against specimens of two important species (*Ixodes ricinus* and *Rhipicephalus sanguineus*). Ticks can transmit a broad spectrum of agents of diseases in cattle or sheep and the use of an effective long lasting acaricide is needed to protect livestock. Four sheep and four young cattle were treated along the vertebral column with 10ml Butox® (deltamethrin) per sheep or 30ml Butox® per cattle. Day 7, 14, 21, and 28 after the treatment, hair was shaved off from the head, ears, the back, belly, and the feet being collected in separate, suitable plastic bags, and transported to the institute, where these hair were brought into close contact with either adult and/or nymph stages of *I. ricinus* and *R. sangui-neus*. As results, strong, acaricidal effects were seen, which varied according to the parasite species, the origin of the hair (e.g., head, leg, etc.) and according to the period after the treatment.

In sheep, the acaricidal effect was noted for the whole period of 28days along the whole body with respect to adults and nymphs of *I. ricinus*, while the acaricidal effects of Deltamethrin were reduced for *R. sanguineus* stages beginning at day 21 after treatment. In cattle, the full acaricidal effect was seen for 21 days in *I. ricinus* stages and for 14 days in *R. sanguineus*, while the acaricidal efficacy became reduced after these periods of full action—beginning at the hair taken from the legs. Only *R. sanguineus* adults did not show any reaction on day 28 after treatment. Besides these acaricidal effects, repellent effects were also noted. Full repellency for both species was seen during the first 14 days in sheep and cattle against *Ixodes* and *Rhipice-phalus*, while the repellency was later reduced, especially in contact with hair from the legs. As conclusion, Deltamethrin, besides its very good effects against biting insects, brings acaricidal as well as repellent effects against ticks, thus protecting the sheep and cattle from transmission of agents of diseases.

3.3. Nuisance flies

Other researchers carried out experiments testing the efficacy of Deltamethrin against nuisance flies of ruminants. More precisely, Franc and Cadierques in 1994, applied the pour on formu-lation of 0.75 % Deltamethrin to cattle on 6 farms in southwest France (3 treated, 3 control). Ten ml per 100 Kg body weight were applied to the backs of 77 cattle (with a max. of 30 ml) and adequate control of hornfles (*Haematobia irritans* and *Hippobosca equina* (with a 95% or better) was achieved during 10 weeks. The protection against non biting flies *Musca autumna-lis* was better than 75 % during 6 weeks and better than 50 % during the next 14 weeks in 2 farms. In the other farm the number of *Musca autumnalis* fell by 75.5% only during the first week.

3.4. Mange

Finally, in the international literature, in 2001, exists a publication of Khalaf-Allah and El-Bablly, who evaluated the effect of Deltamethrin for control of sarcoptic mange in naturally infested cattle. The infested calves (28 calves with 8-16 months of age) were randomly allocated into two groups each consisting of 14 animals. The first group was sprayed with Deltamethrin at the concentration recommended by the manufacturer using a motor sprayer, while the second group was left untreated and served as control. Besides this, 14 healthy calves at the same farm were used to compare between them and the infested calves for the haematological and biochemical parameters. Before application of the acaricide, mange lesions were carefully scraped so as to remove the scales and crusts under which sarcoptic mites are hidden. Skin scrapings were taken from the affected lesions and mites were identified at day (0) and at weekly intervals post-treatment.

The results revealed that Deltamethrin provided a high level of sarcoptic mange control which lasted up to 42 days post-treatment. The mean haematological values of RBCs, Hb, PCV were significantly lower in mange infested calves than that of control, whereas the mean WBCs was significantly higher in infested animals. As well, the mean biochemical parameters estimated in mange infested animals were significantly lower than that of controls. The mean values of the haematological and biochemical parameters in infested animals were restored and nearly returned to its normal levels one month post-treatment.

3.5. Fleas

Herein follow our results from a pilot study which was carried out in order to evaluate under field practise the effect of Deltamethrin (Butox®, MSD) against fleas infesting small ruminants in Greece.

Fleas pose a significant problem in dairy sheep and goat farms of the country, since they attack not only animals but farmers as well (Figure 2). There are several papers in the international literature, regarding flea infestation of livestock in many countries around the Mediterranean basin and elsewhere (Ethiopia, Greece, Israel, Libya, Morocco etc). In all the cases of severe infestation, fleas cause, additionally to nuisance, high mortality, morbidity and disease transfer.

Control is difficult, because fleas spend much time off the host. Furthermore, insecticide residues in milk, when treatment is applied during milk production, are a restraining factor. The great advantage of deltamethrin (Butox®, MSD) is the very short withdrawal period in milk (12 hours) making treatment against ectoparasites practically possible at any time of animal production. Very limited information, to our knowledge, exists in the scientific literature, regarding flea control using insecticides on livestock.

Twenty (15 goat and 5 sheep) farms were identified and Deltamethrin (Butox®, MSD) was applied to all animals at the recommended dose rate. Herds/flocks consisted of 100-200 head of local dairy breeds. Information was collected regarding the management system of the farms, particularly on manure handling (Figure 3).

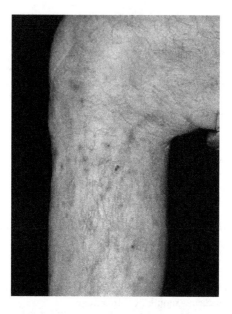

Figure 2. Reaction on human leg after flea feeding

Figure 3. A typical farm environment favouring flea reproduction

Animals within each farm were randomly inspected and fleas, if present, were counted every week for a minimum period of one month (Figure 4). Controls, untreated animals, were not used (accepted by the WAAVP guidelines) for both ethical reasons and because the aim of the study was to eliminate fleas from the farm premises. Practically no fleas were found during the post-treatment period. In more details, the mean (±sd) number of fleas before and after the Deltamethrin treatment were 104.5 (±12.6) and 3.6 (±2.3 fleas), respectively. The overall success of flea control was >96.6%. The main flea species identified was *Ctenocephalides felis*, which is known to be very common and widespread. These results offer a sustainable approach to flea control in Greece due to the long protection period and if combined with hygienic treatment of the farm premises, may contribute significantly to flea control.

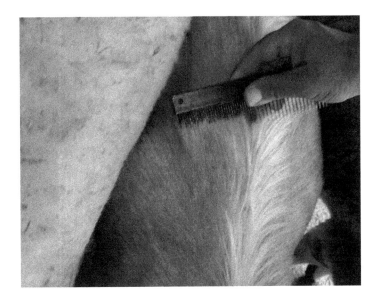

Figure 4. The presence of fleas was inspected using a comb

4. Concluding remarks

In conclusion, Deltamethrin (Butox®, MSD) can be successfully used for farm animal protection in control programmes against many arthropods with important vector-borne or nuisance capacity, including midges, ticks, flies and fleas. Effective control of ectoparasites is of major significance, not only for increased animal production and welfare, but for the public health protection as well.

Author details

Papadopoulos Elias

Laboratory of Parasitology and Parasitic Diseases, Faculty of Veterinary Medicine, Aristotle University of Thessaloniki, Greece

References

[1] Anadon, A, Martinez-larranaga, M. R, & Martinez, M. A. (2009). Use and abuse of pyrethrins and synthetic pyrethroids in veterinary medicine. The Veterinary Journal, 182, 7-20.

[2] Franc, M, & Cadierques, M. C. (1994). Deltamethrin pour-on anti cattle fly trial. Revue Medicale Veterinaire, 145, 337-342.

[3] Khalaf-allah, S. S, & Bablly, M. A. (2001). The efficacy of deltamthrin (a synthetic pyrethroid acaricide) for control of sarcoptic mange in cattle farms. Journal of Egyptian Veterinary Medical Association , 61, 51-59.

[4] Lehmann, K, Werner, D, Hoffmann, B, & Kampen, H. (2012). PCR identification of culicoid biting midges (Diptera, Ceratopogonidae) of the *Obsoletus* complex including putative vectors of bluetongue and Schmallenberg viruses. Parasite Vectors 5, 213.

[5] Mehlhorn, H, Schumacher, B, Jatzlau, A, Abdel-ghaffar, F, Al-rasheid, K, Klimpel, S, & Pohle, H. (2011). Efficacy of deltamethrin (Butox® 7.5 pour on) against nymphs and adults of ticks (*Ixodes ricinus, Rhipicephalus sanguineus*) in treated hair of cattle and sheep. Parasitology Research , 108, 963-971.

[6] Papadopoulos, E, & Farmakis, G. (2012). The use of deltamethrin on farm animals: Our experience on flea control of small ruminants. European Multicolloquium of Parasitology (EMOP XI), 25-29/7/2012 Cluj-Napoca, Romania, , 285-286.

[7] Ross, M. K. (2011). Pyrethroids. Encyclopedia of Environmental Health, , 702-708.

[8] Schmahl, G, Klimpel, S, Walldorf, V, Al-quraishy, S, Schumacher, B, Jatzlau, A, & Mehlhorn, H. (2008). Pilot study on deltamethrin treatment (Butox® 7.5, Versatrine®) of cattle and sheep against midges (*Culicoides* species, Ceratopogonidae). Parasitology Research DOIs00436-008-1260-5.

[9] Schmahl, G, Mehlhorn, H, Abdel-ghaffar, F, Al-rasheid, K, Schumacher, B, Jatzlau, A, & Pohle, H. (2009). Does rain reduce the efficacy of Butox 7.5 pour on (deltamethrin) against biting midges (*Culicoides* specimens)? Parasitology Research , 105, 1763-1765.

[10] Taylor, M. A, Coop, R. L, & Wall, R. L. (2007). Veterinary Parasitology. Blackwell Publishing, Oxford.

Chapter 8

Advances in Insecticide Tools and Tactics for Protecting Conifers from Bark Beetle Attack in the Western United States

Christopher J. Fettig, Donald M. Grosman and
A. Steven Munson

Additional information is available at the end of the chapter

1. Introduction

Bark beetles (Coleoptera: Curculionidae, Scolytinae), a large and diverse group of insects consisting of ~550 species in North America and >6,000 species worldwide, are primary disturbance agents in coniferous forests of the western U.S. Population levels of a number of species (<1%) oscillate periodically, often reaching densities that result in extensive tree mortality when favorable climatic (e.g., droughts) and forest conditions (e.g., dense stands of susceptible hosts) coincide (Table 1). The genera *Dendroctonus*, *Ips* and *Scolytus* are well recognized in this regard. In recent decades, billions of conifers across millions of hectares have been killed by native bark beetles in forests ranging from Alaska to New Mexico, and several recent outbreaks are considered the largest and most severe in recorded history.

Host selection and colonization behavior by bark beetles are complex processes. Following initial attacks and subsequent mating, adults lay eggs in the phloem and larvae excavate feeding tunnels in this tissue and/or the outer bark. Depending on the bark beetle species and the location and severity of feeding, among other factors, this process may result in mortality of the host tree. Top-kill and/or branch mortality are not uncommon. Following pupation, adult beetles of the next generation tunnel outward through the bark and initiate flight in search of new hosts. The lifecycle may be repeated once every several years or several times a year depending on the bark beetle species, geographic location and associated climatic conditions. Extensive levels of tree mortality may result in host replacement by other tree species and plant associations, and may impact timber and fiber production, water quality and quantity, fish and wildlife populations, aesthetics, recreation, grazing capacity, real estate values, biodiversity, carbon storage, endangered species and cultural resources.

Common name	Scientific name	Primary host(s)
Arizona fivespined ips	Ips lecontei	Pinus ponderosa
California fivespined ips	I. paraconfusus	P. contorta, P. jeffreyi, P. lambertiana, P. ponderosa
Douglas-fir beetle	Dendroctonus pseudotsugae	Pseudotsuga menziesii
eastern larch beetle	D. simplex	Larix laricina
fir engraver	Scolytus ventralis	Abies concolor, A. grandis, A. magnifica
Jeffrey pine beetle	D. jeffreyi	P. jeffreyi
mountain pine beetle*	D. ponderosae	P. albicaulis, P. contorta, P. flexilis, P. lambertiana, P. monticola, P. ponderosa
northern spruce engraver	I. perturbatus	Picea glauca, Pi. x lutzii
pine engraver	I. pini	P. contorta, P. jeffreyi, P. lambertiana
pinyon ips	I. confusus	P. edulis, P. monophylla
roundheaded pine beetle	D. adjunctus	P. arizonica, P. engelmannii, P. flexilis, P. leiophylla, P. ponderosa, P. strobiformis
southern pine beetle	D. frontalis	P. engelmannii, P. leiophylla, P. ponderosa
spruce beetle*	D. rufipennis	Pi. engelmannii, Pi. glauca, Pi. pungens, Pi. sitchensis
western balsam bark beetle	Dryocoetes confusus	A. lasiocarpa
western pine beetle*	D. brevicomis	P. coulteri, P. ponderosa

*Species for which preventative insecticide treatments have been well studied.

Table 1. Bark beetle species that cause significant amounts of tree mortality in coniferous forests of the western U.S.

While native bark beetles are a natural part of the ecology of forests, the economic and social impacts of outbreaks can be substantial. Several tactics are available to manage bark beetle infestations and to reduce associated levels of tree mortality. While these vary by bark beetle species, current tactics include tree removals that reduce stand density (thinning) and presumably host susceptibility [1]; sanitation harvests [1]; applications of semiochemicals (i.e., chemicals produced by one organism that elicit a response, usually behavioral, in another organism) to protect individual trees or small-scale stands (e.g., <10 ha) [2]; and preventative applications of insecticides to individual trees. The purpose of this chapter is to synthesize information on the efficacy, residual activity, and environmental safety of insecticides commonly used to protect trees from bark beetle attack so that informed, judicious decisions can be made concerning their use.

2. Types and use of preventative applications of insecticides

Preventative applications of insecticides involve topical sprays to the tree bole (bole sprays) or systemic insecticides injected directly into the tree (tree injections) [3]. Systemic insecticides applied to the soil are generally ineffective. In an operational context, only high-value, individual trees growing in unique environments or under unique circumstances are treated. These may include trees in residential (Fig. 1), recreational (e.g., campgrounds) (Fig. 2) or administrative sites. Tree losses in these environments result in undesirable impacts such as reduced shade, screening, aesthetics, and increased fire risk. Dead trees also pose potential hazards to public safety requiring routine inspection, maintenance and eventual removal [4], and property values may be negatively impacted [5]. In addition, trees growing in progeny tests, seed orchards, or those genetically resistant to forest diseases may be considered for preventative treatments, especially if epidemic populations of bark beetles exist in the area. During large-scale outbreaks, hundreds of thousands of trees may be treated annually in the western U.S., however once an outbreak subsides (i.e., generally after one to several years) preventative treatments are often no longer necessary.

Figure 1. Tree mortality attributed to western pine beetle in San Bernardino County, California, U.S. In the wildland urban interface, tree losses pose potential hazards to public safety and costs associated with hazard tree removals can be substantial. Furthermore, property values may be significantly reduced. The value of these trees, cost of removal and loss of aesthetic value often justify the use of insecticides to protect trees from bark beetle attack during an outbreak. Photos: C.J. Fettig, Pacific Southwest Research Station, USDA Forest Service.

Figure 2. Conditions before (left) and after (right) a spruce beetle outbreak impacted the Navajo Lake Campground on the Dixie National Forest, Utah, U.S. Daily use decreased substantially due to reductions in shade, screening and aesthetics associated with mortality and removal of large diameter overstory trees. Photos: A.S. Munson, Forest Health Protection, USDA Forest Service.

Although once common, insecticides are rarely used today for direct or remedial control (i.e., subsequent treatment of previously infested trees or logs to kill developing and/or emerging brood). While remedial applications have been demonstrated to increase mortality of brood in treated hosts, there is limited evidence of any impact to adjacent levels of tree mortality. Furthermore, there are concerns about the effects of remedial treatments on non-target invertebrates, specifically natural enemy communities. Many of these species respond kairomonally to bark beetle pheromones and host volatiles, and their richness increases over time [6], suggesting that the later remedial treatments are applied the more likely non-target organisms will be negatively impacted.

3. Insecticide registrations

Insecticide sales and use in the U.S. are regulated by federal (U.S. Environmental Protection Agency, EPA) and state (e.g., California Department of Pesticide Regulation in California) agencies. Therefore, product availability and use vary by state. EPA regulates all pesticides under broad authority granted in two statutes, (1) the Federal Insecticide, Fungicide, and Rodenticide Act (FIFRA) that requires all pesticides sold or distributed in the U.S. to be registered; and (2) the Federal Food, Drug and Cosmetic Act that requires EPA to set pesticide tolerances for those used in or on food. EPA may authorize limited use of unregistered pesticides or pesticides registered for other uses under certain circumstances. Under Section 5 of FIFRA, EPA may issue experimental use permits that allow for field testing of new pesticides or uses. Section 18 of FIFRA permits the unregistered use of a pesticide in a specific geographic area for a limited time if an emergency pest condition exists. Under Section 24(c) of FIFRA, states may register a new pesticide for any use, or a federally-registered product for an additional use, as long as a "special local need" is demonstrated.

A complete list of active ingredients and products used for protecting trees from bark beetle attack is beyond our scope as availability changes due to cancellations, voluntary with-

draws, non-payment of annual registration maintenance fees, and registration of new products at federal and state levels. Several studies have been published on the efficacy of
various classes, active ingredients, and formulations that are no longer registered [e.g., benzene hexachloride (Lindane®)]. Therefore, we limit much of our discussion to the most commonly used and/or extensively-studied products (Fig. 3). A list of products registered for
protecting trees from bark beetle attack can be obtained online from state regulatory agencies and/or cooperative extension offices, and should be consulted prior to implementing
any treatment. Furthermore, all insecticides registered and sold in the U.S. must carry a label. It is a violation of federal law to use any product inconsistent with its labeling. The label
contains abundant information concerning the safe and appropriate use of insecticides (e.g.,
signal words, first aid and precautionary statements, proper mixing, etc.). For tree protection, it is important to note whether the product is registered for ornamental and/or forest
settings, and to limit applications to appropriate sites using suitable application rates.

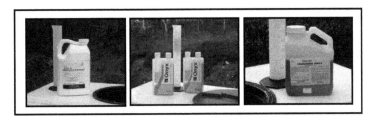

Figure 3. The carbamate carbaryl and pyrethroids bifenthrin and permethrin are commonly used to protect trees from
bark beetle attack in the western U.S. Several formulations are available and effective if properly applied. Residual activity varies with active ingredient, bark beetle species, tree species, geographic location, and associated climatic conditions. Photos: C.J. Fettig, Pacific Southwest Research Station, USDA Forest Service.

4. Experimental designs for evaluating preventative treatments

When evaluating preventative treatments one of three experimental designs is generally
used. Each has its own advantages and disadvantages. Laboratory assays require trapping and/or rearing of live bark beetles for inclusion in experiments. Captured individuals are immediately transported to the laboratory, identified and sorted. Damaged (e.g.,
loss of any appendages), weakened, or beetles not assayed within 48 h after collection
should be discarded. Generally, serial dilutions of each insecticide are prepared, and toxicity is determined in filter paper or topic assays [7]. The life-table method is used to estimate the survival probability of test subjects to different doses of each insecticide [7].
Filter paper assays more closely approximate conditions under which toxicants are encountered by bark beetles during host colonization, especially for products other than
contact insecticides [7], but both methods ignore important environmental factors (e.g.,
temperature, humidity and sunlight) and host tree factors (e.g., architecture) that influ-

ence efficacy. However, results are rapidly obtained with limited risk and loss of scientific infrastructure compared to field studies.

A second design involves field assays in which insecticides are applied to an experimental population of ~25–35 uninfested trees [8]. Trees are often baited with a bark beetle species-specific attractant to increase beetle "pressure" and challenge the treatment following application. Efficacy is based on tree mortality and established statistical parameters [8]. This design is accepted as the standard for evaluating preventative treatments for tree protection in the western U.S., and provides a very conservative test of efficacy [9]. However, it is laborious, time-consuming (i.e., generally efficacy is observed for at least two field seasons) and expensive. Experimental trees may be lost to woodcutting or wildfire, and ≥60% of the untreated control trees must die from bark beetle attack to demonstrate that significant bark beetle pressure exists in the area or the experiment fails and results are inconclusive [8]. Some have argued that the design is perhaps too conservative as under natural conditions aggregation pheromone components would not be released for such extended periods of time as often occurs with baiting. Finally, bark beetles may initiate undesirable infestations near experimental trees as a result of baiting, which may be unacceptable under some circumstances.

The "hanging bolt" assay [10], "small-bolt" assay [11] and similar variants have received limited attention in the western U.S. Typically, insecticides are applied to individual, uninfested trees that are later harvested and cut into bolts for inclusion in laboratory and/or field experiments. Alternatively, freshly-cut bolts may be treated directly in the laboratory. Efficacy is often based on measures of attack density or gallery construction by adult beetles. Compared to [8], these methods allow for rapid acquisition of data; reduced risk of loss to scientific infrasture; and increased probability that a rigorous test will be achieved as bolts are transported to active infestations or brought into the laboratory and exposed to beetles. While these methods account for some host factors (e.g., bark architecture), others such as host defenses and environmental factors are ignored. Furthermore, the hanging bolt and small bolt assays do not provide an estimate of tree mortality, while the effectiveness of any preventative treatment is defined by reductions in tree mortality.

5. Topical applications to the tree bole

Topical applications to protect trees from bark beetle species such as western pine beetle, *Dendroctonus brevicomis* LeConte, mountain pine beetle, *D. ponderosae* Hopkins, and spruce beetle, *D. rufipennis* (Kirby), are applied with ground-based sprayers at high pressure [e.g., ≥2,241 kPa] to the tree bole. Insecticides are applied on all bole surfaces up to a height of ~10.6 to 15.2 m until runoff generally from the root collar to mid-crown (Fig. 4). For engraver beetles, *Ips* spp., that typically colonize smaller diameter hosts branches >5 cm diameter should also be treated. The amount of material (product + water) applied varies with bark and tree architecture, tree size, equipment and applicator, among other factors, but ranges from ~15 to 30 L per tree under most circumstances [12-14]. Application efficiency, the percentage of material applied that is retained on trees, ranges from ~80 to 90% [14].

Figure 4. A common method of protecting trees from bark beetle attack is to saturate all surfaces of the tree bole using a ground-based sprayer at high pressure. Photos: C.J. Fettig, Pacific Southwest Research Station, USDA Forest Service.

Bole sprays are typically applied in late spring prior to initiation of the adult flight period for the target bark beetle species. However, bole sprays require transporting sprayers and other large equipment, which can be problematic in high-elevation forests where snow drifts and poor road conditions often limit access. Additionally, many recreation sites (e.g., campgrounds) where bole sprays are frequently applied occur near intermittent or ephemeral streams that are associated with spring runoff, limiting applications in late spring due to restrictions concerning the use of no-spray buffers to protect non-target aquatic organisms. For these and other reasons, researchers are evaluating alternative timings of bole sprays and less laborious delivery methods.

5.1. Carbaryl

Carbaryl is an acetylcholinesterase inhibitor that prevents the cholinesterase enzyme from breaking down acetylcholine, increasing both the level and duration of action of the neurotransmitter acetylcholine, which leads to rapid twitching, paralysis and ultimately death. Carbaryl is considered essentially nontoxic to birds, moderately toxic to mammals, fish and amphibians, and highly toxic to honey bees, *Apis mellifera* L., and several aquatic insects [15]. However, carbaryl is reported to pose little or no threat to warm-blooded animals. Several experts report that carbaryl is still the most effective, economically-viable, and ecologically-compatible insecticide available for protecting individual trees from mortality due to bark beetle attack in the western U.S. [9,16]. Today, carbaryl (e.g., Sevin® SL and Sevin® XLR Plus, among others) is commonly used to protect trees from bark beetle attack, and is the most-extensively studied active ingredient registered for use. Failures in efficacy are rare and typically associated with inadequate coverage, improper mixing (e.g., using an alkaline water source with pH >8) [17] or inaccurate mixing resulting in solutions of reduced concentration, improper storage, and/or improper timing (e.g., applying treatments to trees already successfully attacked by bark beetles).

Mountain and western pine beetles. Several rates and formulations of carbaryl have been evaluated, and most research indicates two field seasons of protection can be expected with a single application. The effectiveness of 1.0% and 2.0% Sevimol® was demonstrated in the early 1980s [18-22]. This and other research [23-24] led to the registration of 2.0% Sevimol® as a preventative spray, which was voluntarily canceled in 2006. [22] evaluated the efficacy of 0.5%, 1.0% and 2.0% Sevimol® and Sevin® XLR and found all concentrations and formulations were effective for protecting lodgepole pine, *P. contorta* Dougl. ex Loud., from mortality due to mountain pine beetle attack for one year. The 1.0% and 2.0% rates were efficacious for two years. [9] reported 2.0% Sevin® SL protected ponderosa pine, *Pinus ponderosa* Dougl. ex Laws., from western pine beetle attack in California; ponderosa pine from mountain pine beetle attack in South Dakota; and lodgepole pine from mountain pine beetle attack in Montana (two separate studies) for two field seasons. Similar results have been obtained elsewhere [12]. Ongoing research is evaluating the efficacy of fall versus spring applications of 2.0% Sevin® SL for protecting lodgepole pine from mountain pine beetle attack in Wyoming. Both treatments provided 100% tree protection during the first field season while 93% mortality was observed in the untreated control (C.J.F. and A.S.M., unpublished data). A similar study is being conducted for mountain pine beetle in ponderosa pine in Idaho.

Southern pine beetle. Southern pine beetle, *D. frontalis* Zimmerman, occurs in a generally continuous distribution across the southern U.S., roughly coinciding with the distribution of loblolly pine, *P. taeda* L. However, southern pine beetle also occurs in portions of Arizona and New Mexico where it colonizes several pine species, and is therefore considered here. While preventative treatments have not been evaluated in western forests, carbaryl is ineffective for protecting loblolly pine from mortality due to southern pine beetle attack in the southern U.S. [25-26]. This was later linked to insecticide tolerance in southern pine beetle associated with an efficient conversion of carbaryl into metabolites, and a rapid rate of excretion [27-29]. Therefore, despite important environmental differences between the southern and western U.S., carbaryl is regarded as ineffective for preventing southern pine beetle attacks and subsequent tree mortality in the western U.S. [30].

Spruce beetle. Most research suggests that three field seasons of protection can be expected with a single application of carbaryl. In south-central Alaska, [31] reported that 1.0% and 2.0% Sevin® SL protected white spruce, *Picea glauca* (Moench) Voss, and Lutz spruce, *P. glauca* X *lutzii* Little, from attack by spruce beetle for three field seasons, despite early work indicating carbaryl was ineffective in topical assays [32]. One and 2.0% Sevimol® were effective for protecting Engelmann spruce, *P. engelmannii* Parry ex. Engelm., from spruce beetle attack for two field seasons in Utah [33], which agrees with results from [9] for 2.0% Sevin® SL. However, the two latter studies were concluded after two field seasons. In the case of [9], all Sevin® SL-treated trees were alive at the end of the study.

Red turpentine beetle. Red turpentine beetle, *D. valens* LeConte, usually colonizes the basal portions of stressed, weakened, or dead and dying trees. Therefore, the species is not considered an important source of tree mortality in the western U.S., and limited work has occurred regarding the development of tree protection tools. [34] reported that 2.0% Sevin® XLR and 4.0% Sevimol® 4 were effective for protecting ponderosa pine in California. Several formulations of carbaryl are effective for protecting Monterey pine, *P. radiata* D. Don, [35], but residual activity is generally short-lived (<1 yr).

Engraver beetles. A single application of 2.0% Sevin® SL was effective for protecting single-leaf pinyon, *P. monophylla* Torr. & Frem., from mortality due to pinyon ips, *I. confusus* (LeConte), for two field seasons in Nevada [9]. A similar study in pinyon pine, *P. edulis* Engelm., on the Southern Ute Reservation in Colorado found 2.0% Sevin® SL was efficacious for one field season, but bark beetle pressure was insufficient the second year of the study to make definitive conclusions regarding efficacy [9]. [9] also evaluated the efficacy of 2.0% Sevin® SL for protecting ponderosa pine from pine engraver, *I. pini* (Say), but very few trees were attacked during the experiment. Approximately one year later, trees in this study were harvested and cut into bolts that were then laid on the ground in areas containing slash piles infested with pine engraver, sixspined ips, *I. calligraphus* (Germar), and Arizona five-spined ips, *I. lecontei* Swain [13]. From this and related research, the authors concluded 1.0% and 2.0% Sevin® SL were effective for protecting ponderosa pine from engraver beetle attacks for one entire flight season in Arizona. [36] reached similar conclusions for 2.0% Sevin® 80 WSP for a complex of engraver beetles, including sixspined ips, that colonize loblolly pine in the southeastern U.S.

5.2. Pyrethroids

Pyrethroids are synthesized from petroleum-based chemicals and related to the potent insecticidal properties of flowering plants in the genus *Chrysanthemum*. They are axonic poisons and cause paralysis by keeping the sodium channels open in the neuronal membranes [37]. First generation pyrethroids were developed in the 1960s, but are unstable in sunlight. By the mid-1970s, a second generation was developed (e.g., permethrin, cypermethrin and deltamethrin) that were more resistant to photodegradation, but have substantially higher mammalian toxicities. Third generation pyrethroids (e.g., bifenthrin, cyfluthrin and lambda-cyhalothrin) have even greater photostability and insecticidal activity compared to previous generations. Pyrethroids are one of the least acutely toxic insecticides to mammals, essentially nontoxic to birds, but are highly toxic to fish, amphibians and honey bees [38]. Today, permethrin (e.g., Astro® and Dragnet®, among others) and bifenthrin (e.g., Onyx™) are commonly used to protect trees from bark beetle attack, and following carbaryl are the most-extensively studied active ingredients registered for use.

Mountain and western pine beetles. Several active ingredients and formulations of pyrethroids have been evaluated as preventative treatments, and most research indicates at least one field season of protection can be expected with a single application. [8] evaluated 0.1%, 0.2% and 0.4% permethrin (Pounce®) for protecting ponderosa pine from mortality due to western pine beetle attack, and reported that 0.2% and 0.4% provided control for four months. Permethrin plus-C (Masterline®), a unique formulation containing methyl cellulose (i.e., "plus-C") thought to increase efficacy and stability by reducing photo-, chemical- and biological-degradation of the permethrin molecule, exhibits efficacy similar to that of other formulations of permethrin [12]. [39] examined several rates of esfenvalerate (Asana® XL) and cyfluthrin (Tempo® 20 WP) as preventative treatments. In California, 0.025% and 0.05%

Asana® XL protected ponderosa pine for western pine beetle attack for one field season, but not a second. In Montana, 0.006% and 0.012% Asana® XL were ineffective for protecting lodgepole pine from mountain pine beetle, but 0.025% was effective for one field season. Tempo® 20 WP applied at 0.025% provided protection of ponderosa pine from western pine beetle for one field season in Idaho, but not California [39]. Surprisingly, 0.025%, 0.05% and 0.1% Tempo® 20 WP were effective for protecting lodgepole pine from mountain pine beetle attack for two field seasons [39]. [9] evaluated 0.03%, 0.06% and 0.12% bifenthrin (Onyx™) reporting at minimum one field season of protection for mountain pine beetle in lodgepole pine and two field seasons of protection for western pine beetle in ponderosa pine. This study and related research led to the registration of 0.06% Onyx™ as a preventative spray in the mid-2000s. [40] reported 0.06% Onyx™ failed to provide three field seasons of protection for western pine beetle in ponderosa pine, confirming Onyx™ is only effective for two field seasons in that system.

Southern pine beetle. While limited research has occurred, permethrin (Astro®) appears to have longer residual activity than bifenthrin (Onyx™) at least in small-bolt assays [11].

Spruce beetle. Most research suggests that at least one field season of protection can be expected. [9] reported 0.03%, 0.06% and 0.12% bifenthrin (Onyx™) would likely provide protection for two field seasons in Utah. However, 0.025% cyfluthrin (Tempo® 2) and 0.025% and 0.05% esfenvalerate (Asana® XL) only provided one field season of protection in Utah [33]. Protection of Lutz spruce in Alaska is possible for two field seasons with a single application of 0.25% permethrin (formulation unreported) [41].

Red turpentine beetle. [35] reported 0.5% permethrin (Dragnet®) was effective for protecting Monterey pine, and that it had longer residual activity than carbaryl. [34] reported 0.1%, 0.2% and 0.4% permethrin (formulation unreported) were ineffective for protecting ponderosa pine from red turpentine beetle.

Engraver beetles. Most research suggests that at least one field season of protection can be expected with a single application; however, [9] reported 0.03%, 0.06% and 0.12% bifenthrin (Onyx™) protected single-leaf pinyon from pinyon ips for two field seasons in Nevada. A similar study on the Southern Ute Reservation in Colorado found 0.12% Onyx™ protected pinyon pine for one field season, but bark beetle pressure was insufficient the second year of the study to make conclusions regarding efficacy at that rate. Both 0.03% and 0.06% Onyx™ were ineffective [9]. [13] reported that 0.19% permethrin plus-C (Masterline®) and 0.06% bifenthrin (Onyx™) were effective for protecting ponderosa pine bolts from engraver beetle attack in Arizona. [36] reported 0.06% bifenthrin (Onyx™) significantly reduced colonization of trees by bark and woodboring beetles, including sixspined ips, in the southeastern U.S.

6. Systemic injections to the tree bole

Researchers attempting to find safer, more portable and longer-lasting alternatives to bole sprays have evaluated the effectiveness of injecting small quantities of systemic insecticides directly into the lower bole. Early work indicated that several methods, active ingredients and formulations were ineffective [e.g., 13,42-44]. In recent years, the efficacy of phloem-mobile active ingredients injected with pressurized systems (e.g., Sidewinder® Tree Injector, Tree I.V. micro infusion® and Wedgle® Direct-Inject™) capable of maintaining >275 kPA have been evaluated for engraver beetles, mountain pine beetle, southern pine beetle, spruce beetle, and western pine beetle (Fig. 5). These systems push adequate volumes of product (i.e., generally less than several hundred ml for even large trees) into the small vesicles of the sapwood [45]. Applications take <15 minutes per tree under most circumstances. Following injection, the product is transported throughout the tree to the target tissue (i.e., the phloem where bark beetle feeding occurs). Injections can be applied at any time of year when the tree is actively translocating, but time is needed to allow for full distribution of the active ingredient within the tree prior to the tree being attacked by bark beetles. Under optimal conditions (e.g., adequate soil moisture, moderate temperatures and good overall tree health) this takes ~4 weeks [46], but may take much longer, particularly in high-elevation forests. Tree injections represent essentially closed systems that eliminate drift, and reduce non-target effects and applicator exposure, but efficacy is often less than that observed for bole sprays in high-elevation forests [40]. Significant advancements in the development of this technology have been made in recent years, but tree injections are still rarely used in comparison to bole sprays in the western U.S. With the advent of designer formulations of insecticides specific for tree injection, we suspect that tree injections will become a more common tool for protecting trees from bark beetle attack in the near future, particularly in areas where bole sprays are not practical (e.g., along property lines or within no-spray buffers).

Figure 5. Experimental injections of emamectin benzoate for protecting trees from western pine beetle attack in Calaveras County, California, U.S. (left), and mountain pine beetle attack in the Uinta-Wasatch-Cache National Forest, Utah, U.S. (right). Photos: C.J. Fettig, Pacific Southwest Research Station, USDA Forest Service (left) and D.M. Grosman, Texas A&M Forest Service (right).

6.1. Emamectin benzoate

Emamectin benzoate is a macrycyclic lactone derived from avermectin B1 (= abamectin) by fermentation of the soil actinomycete *Streptomyces avermitilis* that disrupts neurotransmitters causing irreversible paralysis. Emamectin benzoate is highly toxic to fish and honey bees, and very highly toxic to aquatic invertebrates. It is highly toxic to mammals and birds as well on an acute oral basis, but is dermally benign to mammals. In recent years, emamectin benzoate has received the most attention among systemic injections for protecting trees from bark beetle attack in the western U.S. [40].

Mountain and western pine beetles. [40] evaluated an experimental formulation of 4.0% emamectin benzoate mixed 1:1 with methanol for protecting ponderosa pine from mortality due to western pine beetle attack in California. Results of this study indicate three field seasons of protection can be expected with a single application. To our knowledge, this was the first demonstration of a successful application of a systemic insecticide for protecting individual trees from mortality due to bark beetle attack in the western U.S. This and other research led to the registration of emamectin benzoate (TREE-age™) in 2010 for protecting individual trees from bark beetle attack.

The experimental formulation of emamectin benzoate was ineffective for protecting lodgepole pine from mountain pine beetle attack in Idaho [40], which agrees with field studies conducted in British Columbia and Colorado (D.M.G., unpublished data). Site conditions such as ambient temperatures, soil temperatures and soil moistures may help explain the lack of efficacy observed in these studies as these factors may slow product uptake and translocation within trees in high-elevation forests [40]. As such, failures for protecting lodgepole pine from mountain pine beetle attack were initially attributed to inadequate distribution of the active ingredient following injections made ~5 weeks prior to trees coming under attack by mountain pine beetle [40]. The authors commented that injecting trees in the fall and/or increasing the number of injection points per tree could perhaps increase efficacy. Currently, spring and fall applications of TREE-age™ are being evaluated for protecting lodgepole pine from mortality due to mountain pine beetle attack in Utah. Results for fall treatments are very promising (Table 2).

Southern pine beetle. Several studies have evaluated the efficacy of emamectin benzoate for protecting loblolly pine from mortality due to southern pine beetle attack in the southern U.S. [47, D.G.M., unpublished data]. Most have demonstrated a reduction in tree mortality, but few trees were attacked in the untreated controls, presumable due to low population levels.

Spruce beetle. An experimental formulation of 4.0% emamectin benzoate injected in late August was ineffective for protecting Engelmann spruce from mortality due to spruce beetle attack in Utah [40]. However, the commercial formulation TREE-age™ has yet to be evaluated. Studies are planned to evaluate alternative timings of injection of TREE-age™ (i.e., early summer versus late summer) and the number and position of the injection ports in trees, both of which are thought to influence efficacy [40].

Engraver beetles. Several studies have reported that emamectin benzoate is effective for preventing engraver beetle attacks, including sixspined ips, for at least two years in Texas [46, D.M.G., unpublished data].

Treatment[a]	Rate[b] (/2.54 cm dbh)	Percent mortality[c]
Spring injection	10 ml	33%
Fall injection	10 ml	0%
Untreated control (yr 1)	-	80%
Untreated control (yr 2)	-	60%

[a] Injections occurred in spring (i.e., June, ~1 month prior to peak mountain pine beetle that year) and fall (i.e., September, ~10 months prior to peak mountain pine beetle flight the following year).

[b] dbh = diameter at breast height (1.37 m in height).

[c] Based on presence or absence of crown fade in September 2011. Data obtained from Fettig et al. (unpublished data).

Table 2. The effectiveness of injections of emamectin benzoate (TREE-age™) into the lower bole of lodgepole pine for reducing levels of tree mortality due to mountain pine beetle attack, Uinta-Wasatch-Cache National Forest, Utah, U.S., 2009-2011.

6.2. Abamectin

Abamectin (= avermectin B1) is a natural fermentation product of the soil actinomycete *Streptomyces a. vermitilis*. Like emamectin benzoate, abamectin acts on insects by interfering with neural and neuromuscular transmission. Abamectin is relatively non-toxic to birds, but highly toxic to fish, aquatic invertebrates and honeybees. Most formulated products are of low toxicity to mammals. Ongoing studies indicate Abacide™ 2 is effective for protecting lodgepole pine from mortality due to mountain pine beetle attack in Utah for at least one field season (C.J.F. et al., unpublished data). Similarly, efficacy has been demonstrated for a complex of engraver beetles, including sixspined ips, for three field seasons in Texas (D.G.M., unpublished data). A request to add mountain pine beetle and engraver beetles to the label for Abacide™ 2 may be forthcoming.

6.3. Fipronil

Fipronil is a phenyl pyrazole that disrupts the insect central nervous system by blocking the passage of chloride ions through the gamma-aminobutyric acid (GABA) receptor and gluta-mate-gated chloride channels. This results in hyperexcitation of contaminated nerves and muscles and ultimately death. Fipronil is of low to moderate toxicity to mammals, highly toxic to fish, aquatic invertebrates, honeybees and upland game birds, but is practically non-toxic to waterfowl and other bird species. Fipronil reduced levels of tree mortality due to engraver beetles, including sixspined ips, on stressed trees in Texas [46]. However, fipronil is ineffective for protecting loblolly pine from southern pine beetle [47] and Engelmann spruce from spruce beetle [40,48]. While results are inconclusive [40, 48], fipronil does not appear effective from reducing levels of lodgepole pine mortality due to mountain pine bee-tle attack in Utah or ponderosa pine mortality due to western pine beetle attack in Califor-nia. Thus, registration is not being pursued at this time.

7. Environmental concerns

Most data on the deposition, toxicity, and environmental fate of insecticides in western for-ests come from aerial applications to control tree defoliators, and therefore are of limited ap-plicability to bole sprays or tree injections used to protect trees from bark beetle attack. [49] studied the effects of lindane, chlorpyrifos and carbaryl on a California pine forest soil ar-thropod community by spraying normal levels of insecticide, and levels five times greater than would be operationally used to protect trees from bark beetle attack. The authors con-cluded carbaryl was least disruptive to the soil arthropod community [49]. Persistence and movement of 2.0% carbaryl within soils of wet and dry sites has been evaluated [50]. The highest concentrations of carbaryl were detected within the uppermost soil layers (upper 2.54 cm), with levels exceeding 20 ppm 90 d after application on most sites [50].

Carbaryl is relatively nontoxic to *Enoclerus lecontei* (Wolcott) [51] and *E. sphegeus* (F.) [52], and less toxic than either lindane or chlorpyrifos to *Temnochila chlorodia* (Mannerheim) [51], common predators of bark beetles in the western U.S. [32] measured the remedial efficacy of 0.25%, 0.5%, 1.0%, and 2.0% chlorpyrifos (Dursban®), fenitrothion (Sumithion®) and perme-thrin (Pounce®) on emerged and nonemerged predators and parasites of spruce beetle in Alaska. Two percent Pounce® had the least impact on emerged natural enemies while Durs-ban® and Sumithion® had the greatest impacts. In many cases, the lowest concentrations re-sulted in the highest mortality of emerged parasites and predators (74-94% mortality), but lowest mortality of nonemerged individuals. The authors attributed this to higher concen-trations resulting in prolong emergence [32]. Mortality of nonmerged parasites and preda-tors was <45% for all active ingredients and concentrations, except 2.0% chlorpyrifos [32].

Werner and Hilgert [53] monitored permethrin levels in a freshwater stream adjacent to Lutz spruce that were treated with 0.5% permethrin (Pounce®) to prevent spruce beetle at-tack. Treatments occurred within 5 m of the stream. Maximum residue levels ranged from 0.05 ± 0.01 ppb 5 h after treatment to 0.14 ± 0.03 ppb 8-11 h after treatment, declining to 0.02 ± 0.01 ppb after 14 h. Levels of permethrin in standing pools near the stream were 0.01 ± 0.01 ppb. Numbers of drifting aquatic invertebrates increased two-fold during treatment and four-fold 3 h after treatment and declined to background levels within 9 h. Trout fry, periph-yton and benthic invertebrates were unaffected [53].

Two studies have been published on the amount of drift resulting from carbaryl applica-tions to protect trees from bark beetle attack. In the early 1980s, [54] used spectrophotofluor-ometry to analyze ground deposition from the base of the ponderosa pine to 12 m from the bole in California. In a more recent study, [14] used high performance liquid chromatogra-phy (HPLC) to evaluate ground deposition occurring at four distances from the tree bole (7.6, 15.2, 22.9 and 38.1 m) during conventional spray applications for protecting individual lodgepole pine from mountain pine beetle attack, and Engelmann spruce from spruce beetle attack. Despite substantial differences in these methods (i.e., spectrophotofluorometry limits detection of finer particle sizes that are accounted for with HPLC), they yielded some similar results. For example, [14] reported application efficiencies of 80.9% to 87.2%, while [54] reported values of >80%. Furthermore, [14] found no significant difference in the amount of drift occurring between lodgepole pine and Engelmann spruce at any distance from the tree bole despite differences in application rate and pressure, while [54] reported drift was similar between two methods applied at 276 kPa and 2930 kPa. However, [14] reported higher levels of ground deposition further away from the tree bole, which is expected given use of HPLC, a more sensitive method of detection.

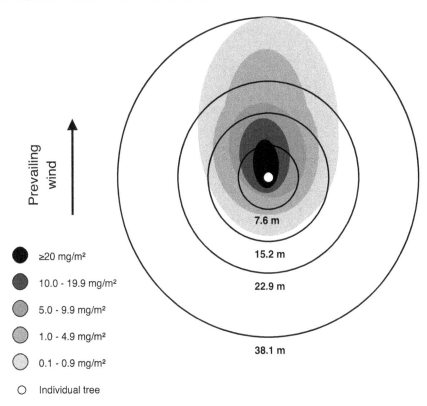

Figure 6. Average drift following experimental applications of carbaryl to protect trees from bark beetle attack, Uinta-Wasatch-Cache National Forest, Utah, U.S. Data obtained from Fettig et al. (2008). Wind speed was correlated with drift up to 22.9 m from the tree bole, and direction largely influenced the direction of prevailing drift. For example, while deposition is detected at 38.1 m on the leeward side of treated trees (maximum wind speeds averaged 3.5 km/h), drift is undetectable less than half that distance on the windward side. Less drift is expected in dense forest stands due to reduced wind speeds and interception by foliage. Studies show no-spray buffers will ensure that adjacent aquatic and terrestrial environments are protected from negative impacts.

Fettig et al. [14] reported mean deposition values from 0.04 ± 0.02 mg carbaryl/m^2 at 38.1 m to 13.30 ± 2.54 mg carbaryl/m^2 at 7.6 m. Overall, distance from the tree bole significantly affected the amount of deposition. Deposition was greatest 7.6 m from the tree bole and declined quickly thereafter. Approximately 97% of total spray deposition occurred within 15.2 m of the tree bole (Fig. 6). To evaluate the potential risk to aquatic environments, the authors converted mean deposition to mean concentration assuming a water depth of 0.3 m selected to represent the average size of lotic systems, primarily small mountain streams, adjacent to many recreational sites where bole sprays are often applied [14]. No adjustments were made for the degradation of carbaryl by hydrolysis, which is rapid in streams or for dilution by natural flow. Comparisons were made with published toxicology data available for select aquatic organisms. No-spray buffers of 7.6 m are sufficient to protect freshwater fish, amphibians, crustaceans, bivalves and most aquatic insects. In laboratory studies, carbaryl was found to be highly toxic to stoneflies (Plecoptera) and mayflies (Ephemeroptera), which are widely distributed and important food sources for freshwater fishes, but negative impacts in field populations are often short-lived and undetectable several hours after contamination [55]. No-spray buffers >22.9 m appear sufficient to protect the most sensitive aquatic insects such as stoneflies.

An advantage of tree injections is that they can be used on environmentally-sensitive sites as these treatments represent an essentially closed system and therefore little or no contamination occurs outside of the tree. However, following injection residues move within the tree and are frequently detected in the foliage [e.g., 44,56-57], which could pose a risk to decomposers and other soil fauna when needles senesce. This has been shown for imidacloprid in maple [57], but injections of emamectin benzoate in pines appear of little risk. For example, [56] reported emamectin benzoate was not detected in the roots or the surrounding soil, but was present at $0.011-0.025$ µg/g in freshly fallen pine needles. However, levels gradually declined to below detectable thresholds after 2 months [56].

8. Conclusions

The results of the many studies presented in this chapter indicate that preventative applications of insecticides are a viable option for protecting individual trees from mortality due to bark beetle attack. Bole sprays of bifenthrin, carbaryl and permethrin are most commonly used. Several formulations are available and effective if properly applied. Residual activity varies with active ingredient, bark beetle species, tree species and associated climatic conditions, but generally one to three years of protection can be expected with a single application. Recent advances in methods and formulations for individual tree injection are promising, and further research and development is ongoing. We expect the use of tree injections to increase in the future. In general, preventative applications of insecticides pose little threat to adjacent environments, and few negative impacts have been observed. We hope that forest health professionals and other resource managers use this publication and other reports to make informed, judicious decisions concerning the appropriate use of preventative treatments to protect trees from mortality due to bark beetle attack. Additional technical assistance in the U.S. can be obtained from Forest Health Protection (USDA Forest Service) entomologists (www.fs.fed.us/foresthealth/), state forest entomologists, and county extension agents (www.csrees.usda.gov/Extension/). We encourage use of these resources before applying any insecticides to protect trees from bark beetle attack.

Acknowledgements

We thank Stanislav Trdan (University of Ljubljana) and Romana Vukelic and Danijela Duric for their invitation to contribute and guidance during the publishing process. Numerous colleagues from Arborjet Inc., BASF, Bayer ES, Bureau of Land Management (U.S.),FMC Corp., Fruit Grower's Supply Co., Mauget Inc., Nevada Division of Forestry, Sierra Pacific Industries, Southern Ute Reservation, Syngenta Crop Protection, Texas A&M Forest Service, Univar USA Inc., University of Arizona, University of California, University of Georgia, and USDA Forest Service have contributed to the success of much of the research discussed in this chapter. We are thankful for their support and helpful insights.

This publication concerns pesticides. It does not contain recommendations for their use, nor does it imply that the uses discussed here have been registered. All uses of pesticides in the United States must be registered by appropriate State and/or Federal agencies. CAUTION: Pesticides can be injurious to humans, domestic animals, desirable plants, and fish or other wildlife—if they are not handled or applied properly. Follow recommended practices for the disposal of surplus pesticides and their containers.

Author details

Christopher J. Fettig[1], Donald M. Grosman[2] and A. Steven Munson[3]

*Address all correspondence to: cfettig@fs.fed.us

1 Pacific Southwest Research Station, USDA Forest Service, Davis, CA, USA

2 Forest Health, Texas A&M Forest Service, Lufkin, USA

3 Forest Health Protection, USDA Forest Service, Ogden, UT, USA

References

[1] Fettig CJ, Klepzig KD, Billings RF, Munson AS, Nebeker TE, Negrón JF, Nowak JT. The Effectiveness of Vegetation Management Practices for Prevention and Control of Bark Beetle Outbreaks in Coniferous Forests of the Western and Southern United States. Forest Ecology and Management 2007; 238 24-53.

[2] Gillette NE, Munson AS. Semiochemical sabotage: Behavioral chemicals for protection of western conifers from bark beetles. In: Hayes JL, Lundquist JE (compilers). The Western Bark Beetle Research Group: A unique collaboration with Forest Health Protection: proceedings of a Symposium at the 2007 Society of American Foresters Conference, 23-28 October 2007, Portland, OR. PNW-GTR-784. Portland, OR: U.S. Department of Agriculture, Forest Service, Pacific Northwest Research Station; 2009. p85-110.

[3] Munson AS, Steed B, Fettig CJ. Using Insecticides to Protect Individual Conifers from Bark Beetle Attack in the West. Ogden, UT: U.S. Department of Agriculture, Forest Service, Forest Health Protection, Intermountain and Northern Regions; 2011.

[4] Johnson DW. Tree Hazards: Recognition and Reduction in Recreation Sites. R2-TR-1. Lakewood, CO: Department of Agriculture, Forest Service, Rocky Mountain Region; 1981.

[5] McGregor MD, Cole WE. Integrating Management Strategies for the Mountain Pine Beetle With Multiple Resource Management of Lodgepole Pine Forests. INT-GTR-174. Ogden, UT; U.S. Department of Agriculture, Forest Service, Intermountain Forest and Range Experiment Station; 1985.

[6] Stephen FM, Dahlsten DL. The Arrival Sequence of the Arthropod Complex Following Attack by *Dendroctonus brevicomis* (Coleoptera: Scolytidae) in Ponderosa Pine. The Canadian Entomologist 1976; 108 283-304.

[7] Fettig CJ, Hayes CJ, McKelvey SR, Mori SL. Laboratory Assays of Select Candidate Insecticides for Control of *Dendroctonus ponderosae* Hopkins. Pest Management Science 2011; 67 548-555.

[8] Shea PJ, Haverty MI, Hall RW. Effectiveness of Fenitrothion and Permethrin for Protecting Ponderosa Pine from Attack by Western Pine Beetle. Journal of the Georgia Entomological Society 1984; 19 427-433.

[9] Fettig CJ, Allen KK, Borys RR, Christopherson J, Dabney CP, Eager TJ, Gibson KE, Herbertson EG, Long DF, Munson AS, Shea PJ, Smith SL, Haverty MI. Effectiveness of Bifenthrin (Onyx®) and Carbaryl (Sevin SL®) for Protecting Individual, High-value Conifers from Bark Beetle Attack (Coleoptera: Curculionidae: Scolytinae) in the Western United States. Journal of Economic Entomology 2006; 99 1691-1698.

[10] Berisford CW, Brady UE, Mizell RF, Lashomb JH, Fitzpatrick GE, Ragenovich IR, Hastings FL. A Technique for Field Testing Insecticides for Long-term Prevention of Bark Beetle Attack. Journal of Economic Entomology 1980; 73 694-697.

[11] Strom BL, Roton LM. A Small-Bolt Method for Screening Tree Protectants Against Bark Beetles (Coleoptera: Curculionidae). Journal of Entomological Science 2009; 44 297-307.

[12] Fettig CJ, DeGomez TE, Gibson KE, Dabney CP, Borys RR. Effectiveness of Permethrin plus-C (Masterline®) and Carbaryl (Sevin SL®) for Protecting Individual, High-

value Pines from Bark Beetle Attack. Arboriculture and Urban Forestry 2006; 32 247-252.

[13] DeGomez TE, Hayes CJ, Anhold JA, McMillin JD, Clancy KM, Bosu PP. Evaluation of Insecticides for Protecting Southwestern Ponderosa Pines from Attack by Engraver Beetles (Coleoptera: Curculionidae, Scolytinae). Journal of Economic Entomology 2006; 99 393-400.

[14] Fettig CJ, Munson AS, McKelvey SR, Bush PB, Borys RR. Spray Deposition from Ground-based Applications of Carbaryl to Protect Individual Trees from Bark Beetle Attack. Journal of Environmental Quality 2008; 37 1170-1179.

[15] Jones RD, Steeger TM, Behl B. Environmental fate and ecological risk assessment for the re-registration of carbaryl. www.epa.gov/espp/litstatus/effects/carb-riskass.pdf (accessed 20 August 2012).

[16] Hastings FL, Holsten EH, Shea PJ, Werner RA. Carbaryl: A Review of Its Use Against Bark Beetles in Coniferous Forests of North America. Environmental Entomology 2001; 30 803-810.

[17] Dittert LW, Higuchi T. Rates of Hydrolysis of Carbamate and Carbonate Esters in Alkaline Solution. Journal of Pharmaceutical Sciences 1963; 52 852-857.

[18] Hall RW, Shea PJ, Haverty MI. Effectiveness of Carbaryl and Chlorpyriphos for Protecting Ponderosa Pine Trees from Attack by Western Pine Beetle (Coleoptera: Scolytidae). Journal of Economic Entomology 1982; 75 504-508.

[19] Gibson KE, Bennett DD. Effectiveness of Carbaryl in Preventing Attacks on Lodgepole Pine by the Mountain Pine Beetle. Journal of Forestry 1985; 83 109-112.

[20] Haverty MI, Shea PJ, Hall RW. Effective Residual Life of Carbaryl for Protecting Ponderosa Pine from Attack by the Western Pine Beetle (Coleoptera: Scolytidae). Journal of Economic Entomology 1985; 78 197-199.

[21] Page M, Haverty MI, Richmond CE. Residual Activity of Carbaryl Protected Lodgepole Pine Against Mountain Pine Beetle, Dillon, Colorado, 1982 and 1983. PSW-RN-375. U.S. Department of Agriculture, Forest Service, Pacific Southwest Research Station; 1985.

[22] Shea PJ, McGregor MD. A New Formulation and Reduced Rates of Carbaryl for Protecting Lodgepole Pine from Mountain Pine Beetle Attack. Western Journal of Applied Forestry 1987; 2 114-116.

[23] Smith RH, Trostle GC, McCambridge WF. Protective Spray Tests on Three Species of Bark Beetles in the Western United States. Journal of Economic Entomology 1977; 70 119-125.

[24] McCambridge WF. Field Tests of Insecticides to Protect Ponderosa Pine from the Mountain Pine Beetle (Coleoptera: Scolytidae). Journal of Economic Entomology 1982; 75 1080-1082.

[25]   Ragenovich IR, Coster JE. Evaluation of Some Carbamate and Phosphate Insecticides Against Southern Pine Beetle and *Ips* Bark Beetles. Journal of Economic Entomology 1974; 67 763-765.

[26]   Berisford CW, Brady UE, Ragenovich IR. Residue Studies. In: Hastings FL, Costner JE. (eds.) Field and Laboratory Evaluations of Insecticides for Southern Pine Beetle Control. SE-GTR-21. Asheville, NC: U.S. Department of Agriculture, Forest Service, Southeastern Forest Experiment Station; 1981. p17-18.

[27]   Zhong H, Hastings FL, Hain FP, Werner RA. Toxicity of Carbaryl Toward the Southern Pine Beetle in Filter Paper, Bark and Cut Bolt Bioassays. Journal of Entomological Science 1994; 29 247-253.

[28]   Zhong H, Hastings FL, Hain FP, Dauterman WC. Comparison of the Metabolic Fate of Carbaryl-naphthyl-1-14C in Two Beetle Species (Coleoptera: Scolytidae). Journal of Economic Entomology 1995; 88 551-557.

[29]   Zhong H, Hastings FL, Hain FP, Monahan JF. Carbaryl Degradation on Tree Bark as Influenced by Temperature and Humidity. Journal of Economic Entomology 1995; 88 558-563.

[30]   Fettig CJ, Munson AS, McKelvey SR, DeGomez TE. Deposition from Ground-based Sprays of Carbaryl to Protect Individual Trees from Bark Beetle Attack in the Western United States. AZ1493. Tucson, AZ: University of Arizona, College of Agriculture and Life Sciences Bulletin; 2009.

[31]   Werner RA, Hastings FL, Holsten EH, Jones AS. Carbaryl and Lindane Protect White Spruce (*Picea glauca*) from Attack by Spruce Beetle (*Dendroctonus rufipennis*) (Coleoptera: Scolytidae) for Three Growing Seasons. Journal of Economic Entomology 1986; 79 1121-1124.

[32]   Werner RA, Hastings FL, Averill RD. Laboratory and Field Evaluation of Insecticides Against the Spruce Beetle (Coleoptera: Scolytidae) and Parasites and Predators in Alaska. Journal of Economic Entomology 1983; 76 1144-1147.

[33]   Johnson KJ. Effectiveness of carbaryl and pyrethroid insecticides for protection of Engelmann spruce from attack by spruce beetle (Coleoptera: Scolytidae). MS thesis. Utah State University, Logan, UT; 1996.

[34]   Hall RW. Effectiveness of Insecticides for Protecting Ponderosa Pines from Attack by the Red Turpentine Beetle (Coleoptera: Scolytidae). Journal of Economic Entomology 1984; 77 446-448.

[35]   Svihra P. Prevention of Red Turpentine Beetle Attack by Sevimol and Dragnet. Journal of Arboriculture 1995; 23 221-224.

[36]   Burke JL, Hanula JL, Horn S, Audley JP, Gandhi KJK. Efficacy of Two Insecticides for Protecting Loblolly Pines (*Pinus taeda* L.) from Subcortical Beetles (Coleoptera: Curculionidae and Cerambycidae). Pest Management Science 2012; 68 1048-1052.

[37] Miller TA, Salgado VL. The Mode of Action of Pyrethroids on Insects. In: Leahy JP. (ed.) The Pyrethroid Insecticides. London: Taylor & Francis; 1985. p43-97.

[38] Pajares JA, Lanier GN. Pyrethroid Insecticides for Control of European Elm Bark Beetle (Coleoptera: Scolytidae). Journal of Economic Entomology 1989; 82 873-878.

[39] Haverty MI, Shea PJ, Hoffman JT, Wenz JM, Gibson KE. Effectiveness of Esfenvalerate, Cyfluthrin, and Carbaryl in Protecting Individual Lodgepole Pines and Ponderosa Pines From Attack by *Dendroctonus* spp. PSW-RP- 237. Albany, CA: U.S. Department of Agriculture, Forest Service, Pacific Southwest Research Station; 1998.

[40] Grosman DM, Fettig CJ, Jorgensen CL, Munson AS. Effectiveness of Two Systemic Insecticides for Protecting Western Conifers from Mortality Due to Bark Beetle Attack. Western Journal of Applied Forestry 2010; 25 181-185.

[41] Werner RA, Averill RD, Hastings FL, Hilgert JW, Brady UE. Field Evaluation of Fenitrothion, Permethrin, and Chlorpyrifos for Protecting White Spruce Trees from Spruce Beetle (Coleoptera: Scolytidae) Attack in Alaska. Journal of Economic Entomology 1984; 77 995-998.

[42] Duthie-Holt MA, Borden JH. Treatment of Lodgepole Pine Bark With Neem Demonstrates Lack of Repellency or Feeding Deterrency to the Mountain Pine Beetle, *Dendroctonus ponderosae* Hopkins (Coleoptera: Scoytidae). Journal of the Entomological Society of British Columbia 1999; 96 21-24.

[43] Haverty MI, Shea PJ, Wenz JM. 1996. Metasystox-R, Applied in Mauget Injectors, Ineffective in Protecting Individual Ponderosa Pines from Western Pine Beetles. PSW-RN-420. Albany, CA: U.S. Department of Agriculture, Forest Service, Pacific Southwest Research Station; 1996.

[44] Shea PJ, Holsten EH, Hard J. Bole Implantation of Systemic Insecticides Does Not Protect Trees from Spruce Beetle Attack. Western Journal of Applied Forestry 1991; 6 4-7.

[45] Sánchez-Zamora MA, Fernández-Escobar R. Uptake and Distribution of Truck Injections in Conifers. Journal of Arboriculture 2004; 30 73-79.

[46] Grosman DM, Upton, WW. Efficacy of Systemic Insecticides for Protection of Loblolly Pine Against Southern Engraver Beetles (Coleoptera: Curculionidae: Scolytinae) and Wood Borers (Coleoptera: Cerambycidae). Journal of Economic Entomology 2006; 99 94-101.

[47] Grosman DM, Clarke SR, Upton WW. Efficacy of Two Systemic Insecticides Injected Into Loblolly Pine for Protection Against Southern Pine Bark Beetles (Coleoptera: Curculionidae). Journal of Economic Entomology 2009; 102 1062-1069.

[48] Fettig CJ, Munson AS, Jorgensen CL, Grosman DM. Efficacy of Fipronil for Protecting Individual Pines from Mortality Attributed to Attack by Western Pine Beetle and Mountain Pine Beetle (Coleoptera: Curculionidae, Scolytinae). Journal of Entomological Science 2010; 45 296-301.

[49] Hoy JB, Shea PJ. Effects of Lindane, Chlorpyrifos, and Carbaryl on a California Pine
      Forest Soil Arthropod Community. Environmental Entomology 1981; 10 732-740.

[50] Hastings FL, Werner RA, Holsten EH, Shea PJ. The Persistence of Carbaryl Within
      Boreal, Temperate, and Mediterranean Ecosystems. Journal of Economic Entomology
      1998; 91 665-670.

[51] Swezey SL, Page ML, Dahlsten DL. Comparative Toxicity of Lindane, Carbaryl, and
      Chlorpyrifos to the Western Pine Beetle and Two of Its Predators. The Canadian En-
      tomologist 1982; 114 397-401.

[52] Greene LE. Simulated Natural Encounters of the Insecticides, Chlorpyrifos and Car-
      baryl, by Western Pine Beetle Predators *Enoclerus lecontei* and *E. sphegeus* (Coleoptera:
      Cleridae). Environmental Entomology 1983; 12 502-504.

[53] Werner RA, Hilgert JW. Effects of Permethrin on Aquatic Organisms in a Freshwater
      Stream in South-central Alaska. Journal of Economic Entomology 1992; 85 860-864.

[54] Haverty MI, Page M, Shea PJ, Hoy JB, Hall RW. Drift and Worker Exposure Result-
      ing From Two Methods of Applying Insecticides to Pine Bark. Bulletin of Environ-
      mental Contamination and Toxicology 1983; 30 223-228.

[55] Beyers DW, Farmer MS, Sikoski PJ. Effects of Rangeland Aerial Application of Sev-
      in-4-Oil on Fish and Aquatic Invertebrate Drift in the Little Missouri River, North
      Dakota. Archives of Environmental Contamination and Toxicology 1995; 28 27-34.

[56] Takai K, Suzuki T, Kawazu K. Distribution and Persistence of Emamectin Benzoate
      at Efficacious Concentrations in Pine Tissues after Injection of a Liquid Formulation.
      Pest Management Science 2004; 60 42-48.

[57] Kreutzweiser DP, Good KP, Chartrand DT, Scarr TA, Thompson DG. Are Leaves
      that Fall from Imidacloprid-treated Maple Trees to Control Asian Longhorned Bee-
      tles Toxic to Non-target Decomposer Organisms? Journal of Environmental Quality
      2008; 37 639-646.

Biotechnology and Other Advances in Pest Control

Polymeric Nanoparticle-Based Insecticides: A Controlled Release Purpose for Agrochemicals

Bruno Perlatti, Patrícia Luísa de Souza Bergo,
Maria Fátima das Graças Fernandes da Silva,
João Batista Fernandes and Moacir Rossi Forim

Additional information is available at the end of the chapter

1. Introduction

Insects are one of the biggest animal populations with a very successful evolutive history, once they can be found chiefly in all possible environments all over the world, and the number of species and individuals. Their success can be attributed to several important evolutionary aspects like wings, malleable exoskeleton, high reproductive potential, habits diversification, desiccation-resistant eggs and metamorphosis, just to name a few. Some species are especially valuable for humans due to their ability in providing several important goods, such as honey, dyes, lac and silk. On the other hand, many insects are vectors of many diseases, and many others damages crop plantations or wood structures, causing serious health and economic issues.

Among all identified insects, over 500,000 species feed on green leaves. About 75% of them have a restrict diet, eating only a limited range of species, sometimes being even specie specific [1]. This kind of insect brings major concern to the agriculture. Their high selectivity implies in a closer insect attack on crops. It is estimated that about 10,000 insect species are plagues and, compromising the food production, either in the field or after the harvest [2]. It was estimated that somewhere around 14-25% of total agriculture production is lost to pests yet [3].

Agriculture is one of the main pillars of human population increase over the last millenniums, providing mankind with several important commodities such as food, fuel, healthcare and wood. This huge production should feed 7 billion people, and also generate several inputs for many industrial processes and commercial applications. In order to combat the numerous losses that are caused by insects on agriculture, several chemicals have been used to kill them or inhibit their reproduction and feeding habits. Those classes of compounds are collectivity known as insecticides. These molecules are able to interfere in the insect metabolism. They alter is in such a way that the plague cannot feeds on the crop or the harvest or even reproduce anymore. The use of insecticides is described since ancient times, with documents providing evidences as far as in the 16th century BC. The *Ebers Papyrus*, wrote by the Egyptians, reports several chemical and organic substances used against overcome fleas, gnats and biting flies among others [4]. Nowadays, the insecticides are widely employed around the world. Several known substances are extremely effective in controlling or even wiping out almost all important agricultural plagues. This multi-billion-dollar has an estimated production of 2 million metric tons of hundreds of chemical and biological different products, with a budget of a US$35 billion dollars worldwide [5].

Insecticides are used in different ways, based on the physical-chemical characteristics of the each chemical substance, the area that needs to be covered and the target. Typical application of insecticides in crops is made by spraying a solution, emulsion or colloidal suspension containing the active chemical compound, which is made by a vehicle which may be a hand pump, a tractor or even a plane. This mixture is prepared using a liquid as a carrier, usually water, to ensure a homogenous distribution. Other methods for applying insecticides are through foggers or granule baits embedded with the active compound, among others that are less used. However, due to several degradation processes, such as leaching or destruction by light, temperature, microorganism or even water (hydrolysis), only a small amount of these chemical products reaches the target site. In this case, the applied concentrations of these compounds have been much higher than the required. On the other hand, the concentration that reaches its target might be lower than the minimum effective one. In general, depending of the weather and method of application, the amount of applied agrochemicals, as much as 90%, may not reach the target and so do not produce the desired biological response. For this reason, repeated application of pesticides become hence necessary to efficient control of target plagues, which increase the cost and might cause undesirable and serious consequences to the ecosystems, affecting human health [6]. Due to the lack of selectivity, their unrestrained use can also lead to the elimination of the natural enemies, what implies in the fast growth of plague population. Moreover, it often makes the insects resistant to the pesticides.

Another important point that needs attention is the formulation for the application of the insecticide on the crops. There are several different classes of compounds, which sometimes do not match with a simple dilution in water and must be prepared by other means such as powders, emulsions or suspensions. Some kinds of formulations must be handled with more precaution, since it can severely contaminate workers on the field with small airborne solid particles that can be inhaled [7].

The advances in science and technology in the last decades were made in several areas of insecticide usage. It includes either the development of more effective and non-persistent pesticides and new ways of application, which includes controlled release formulations (CRFs). The endeavors are direct towards the successful application of those compounds on crops and their efficacy and availability improvement and reduction of environmental con-tamination and workers exposure [8]. In that line, new types of formulation were devel-oped. One of the most promising is the use of micro and nanotechnology to promote a more efficient assembly of the active compound in a matrix.

2. Application of insecticides nanoformulations

2.1. Nanoemulsions

Casanova *et al.* [9] evaluated the production of a nicotine carboxylate nanoemulsion using a series of fatty acids (C10 – C18) and surfactant. The oil-in-water nanoemulsion showed a monomodal distribution of size, with mean particle sizes of 100nm. The bioactivity of the insecticide formulations was evaluated against adults of *Drosophila melanogaster* by assessing the lethal time 50 (LT_{50}). They observed that the encapsulation efficiency decreased with in-creasing size of the fatty acids tested. The bioactivity followed the same trend, with better bioactivity when the chain length decreased. This would be readily attributed to the higher amount of active compound inside the nanoemulsion. For the smallest fatty acid emulsion used, the capric acid (C10) one, the greatest encapsulation efficiency was observed, but it had the lowest bioactivity. The results were explained in terms of lesser bioavailability of the insecticide in its active form due to increased stability of the organic salt formed between the insecticide and the fatty acid. This experiment highlights the necessity of developing differ-ent kinds of possible assembles between the active compounds and matrix, and extensively studying the interactions in nanoscale formulations, where sometimes nontrivial effects might be unexpectedly observed.

Wang *et al.*[10] developed an assemble of oil-in-water nanoemulsion (O/W) with 30 nm droplets by careful control of experiment conditions, using the neutral surfactant poly(oxy-ethylene) lauryl ether and methyl decanoate to encapsulate highly insoluble β-cypermeth-rin. The dissolution of the insecticide was enhanced. The stability tests were performed by spraying nanoemulsion in a glass slide and observing under polarizing light microscopy. They showed no apparent precipitate in nanoemulsions samples. These results were differ-ent from the ones obtained using a commercial β-cypermethrin formulation, with apparent signs of solid residues after 24 hours. This enhanced stability may be used to decrease the concentration of insecticides in commercial spray applications, without losing efficiency.

2.2. Classical micro and nanoparticles

Allan *et al.* [11] published the first report on a controlled release system of an insecticide through a polymeric encapsulation. Even so, at first the encapsulated systems were not so effective. Problems associated with controlled release and particle stability hindered their practical field application for some decades. In one of the first successful works in the field of pesticides encapsulation, Greene *et al.* [12] used poly (n-alkyl acrylates) (Intelimer®) to produce temperature-sensitive microcapsules of the organophosphate insecticide diazinon.. The active chemical was controlled release by increasing the ambient temperature above

30ºC, which is the melting temperature of the polymer,. Experiments were performed with Banded cucumber beetle *Diabrotica balteata* and Western corn rootworm *Diabrotica virgifera* as target insects at 20ºC and 32ºC, under and above the polymer melting point respectively. Mortality was compared to commercial granular formulation. At lower temperatures, the commercial formulation showed the best mortality. At higher temperatures the activity of the encapsulated formulation was better, showing about 90% of mortality for over 8 weeks. The commercial formulation had indeed lost some of its activity, presumably due to heat degradation.

Latheef *et al.* [13] tested several different polymers such as poly (methyl methacrylate) (PMMA), ethyl cellulose, poly(α-methylstyrene) and cellulose acetate butyrate to produce microcapsules of the insecticide sulprofos. Ethyl cellulose formulations were the only ones that had shown good results against eggs and larvae of the tobacco budworm *Heliothis virescens* in cotton plants. The results were comparable to the ones obtained with the use of an emulsifable-concentrate (EC) commercial formulation of sulprofos.

In other to develop commercial formulation containing microencapsulated cyfluthrin, Arthur[14] evaluated its use against the rice weevil *Sitophilus oryzae* in stored wheat, for a period of 8 months. Survival of beetles was statistically correlated with the concentration of the pyrethroid insecticide in the formulation. The average survival rate was only 12% when 4ppm was used, with constant activity throughout the entire experiment. This evidenced the controlled release of the substance over a long period of time.

In the work carried out by Quaglia *et al.* [15], a hydrophobic waxy prepared through a mixture of di- and triglycerides of PEG esters was used to construct microspheres containing the insecticide carbaryl. Microparticles was obtained with particle size ranging from 16 to 20μm. Controllable release dynamics depended on the amount of gelucire used, Studies of release profiles from the encapsulated formulation showed a lower vertical mobility of the insecticide when compared to a commercial nonencapsulated formation. This suggested that the controlled release profile of the microcapsules may be useful to avoid or minimize ground-water contamination.

Cao *et al.* [16] produced diffusion-controlled microcapsules with diameter ranging from 2 to 20μm with encapsulated acetamiprid, an alkaline and high temperature-sensitive insecticide, using tapioca starch as matrix with urea and sodium borate as additives. The particle showed increased degradation resistance by heat for 60 days, and UV radiation over 48h, with no more than 3% of degradation. This represents less than one tenth when compared to the UV degradation of commercial emulsifable concentrate. Even in those conditions, it was also able to promote controlled liberation of the active compound for up to 10 weeks depending on the formulation used.

In another work with acetamiprid, Takei *et al.* [17]produced microparticles with diameter of 30-150μm using poly-lactide (PLA) as the polymeric matrix. Initial results showed that microspheres containing only PLA did not have a good release kinetic of the active chemical compound from its interior. It is presumably due to their tight structure and high hydrophobicity, which hinders water diffusion and therefore limits the insecticide liberation. The inclusion of poly(ϵ-caprolactone) (PCL) into the matrix in 50-80% weight were analyzed, with formation of microspheres of PLA/PCL blend with 20-120μm of the diameter, showing up to 88,5% of insecticide release in aqueous media over a 48h period.

In contrast to conventional desire to produce compounds with extended residual activity, quick-release microcapsules are demanded in certain areas of agriculture. However, sometimes it is also necessary a quick liberation of the active compound from the matrix after the application. The strong backbone might pose as a problem to effectively deliver. Studies performed by Tsuda *et al.* [18,19] have shown that is possible to assemble "self-bursting" microcapsules that retain its form in water suspension, but easily burst after solvent evaporation. They used the interfacial polymerization method to assemble spherical polyurethane microcapsules containing the insecticide pyriprofixen, obtaining particles with mean diameter of 23μm. The entrapment ratio was 99% for all formulations tested, greatly improving the solubility of the pesticide in water. According to the results, there is a correlation between the wall thickness of the microcapsules and the self-bursting phenomenon. Tuning this property a controlled released can be achieved.

The effectiveness of encapsulated formulations, it is not restricted to extend the residual activity of insecticides, but should also include the overcoming of problems associated with accumulation of recalcitrant organic pollutants that remains in ecosystems in amounts above the Maximum Residual Level (MRL). Therefore, it can be harmful to the environment and to people who might consume the treated crops. For instance, Guan *et al.* [20] encapsulated imidacloprid, a chloro-nicotinyl systemic and broad spectrum insecticide in a mixed sodium alginate/chitosan microparticle through self-assembly layer-by-layer (LbL) methodology. The capsules showed a mean diameter of 7μm. Particles were impregnated with a photocatalyst made of SDS/TiO2/Ag, and the photocatalytic property and the insecticidal activity of the microcapsule was evaluated. Prolonged residual activity of the encapsulated formulation was observed. The toxicity was higher in the *Martianus dermestoides* adult stage compared to the one of pure insecticide. In a field test with soybean [21], the nano-imidacloprid formulation prevented the accumulation of the pesticide on the soybean leaves and soil. The results showed pronounced degradation over 25 days of trials when compared to commercial concentrate formulations, even though the initial concentration of both formulations was equivalent. In this way, regardless the initial effectiveness of the insecticide, safer levels of agrochemicals can be obtained in less time, improving the safety of insecticide application.

2.3. Entomopathogenic microorganisms encapsulated

Besides the chemical compounds, the micro- and nanotechnology have also been developed and applied to microorganisms that need special protection or to improve their solubility in aqueous phase. For instance, Ramírez-Lepe *et al.* [22] developed an aluminium-carboxymethylcellulose microcapsule with photoprotective agents for holding a *Bacillus thuringiensis* serovar *israelensis* (B.t.i.) spore-toxin complex named δ-endotoxin. The protein produced by this gram-positive bacterium during sporulation is extremely toxic to larval stage of some mosquitoes and flies which are vectors for important tropical diseases such as malaria and dengue. The encapsulated formulation was tested for its UV irradiation protective efficiency in laboratory conditions. While the protein in its natural form had lost all of its activity after 24 hours of exposure, encapsulated formulations showed up to 88% of larvae mortality.

In their turn, Tamez-Guerra *et al.* [23] also tested the encapsulation of the spore-toxin of *Bacillus thuringiensis* Berliner, evaluating over 80 formulations of spray-dried microcapsules made of lignin and corn flour with and without photoprotective agents. The best formulations showed improved insecticidal activity in laboratory tests against neonates of European corn borer *Ostrinia nubilalis* when compared to nonencapsulated or commercial formulations of the same endotoxin. In a field test, the microcapsules showed increased residual insecticide activity in cabbage after 7 days against neonates of the cabbage looper *Trichoplusia ni* when compared to commercial formulations.

Very promising results have been obtained by the Agricultural Research Service of the USDA regarding the encapsulation of biopesticides made of species-specific nucleopolyhedroviruses (NPV) isolated from several insects, including celery looper *Anagrapha falcifera* (Tamez-Guerra *et al.*, 2000 [24-26]), alfalfa looper *Autographa californica* [27], codling moth *Cydia pomonella* [28] and fall armyworm *Spodoptera frugiperda* [29]. In these works, formulations were developed using different mixtures of corn flour and lignin, through spray-drying technique to encapsulate the viruses. All results obtained in laboratory and field tests performed have shown improvements in insecticidal activity, resistance to environmental conditions, like rain and UV light exposure, and a prolonged residual activity against pests in field studies. Samples were kept in storage for up to 12 months and maintained their insecticidal activity.

2.4. Novel micro and nanoparticles for bioinseticides

Conventional protocols for encapsulation usually run under relatively high temperatures, which might be inadequate for preserving plant-derived essential oils integrity. Processes which use high pressure instead of temperature can be an alternative for encapsulating these sensible extracts. Varona *et al.* [30,31] developed new methods to produce stable particles of lavandin (*Lavandula hybrida*) essential oil, using polyethylene glycol 9000 (PEG9000) or n-octenyl succinic (OSA) modified starches as the shell material. The methods for preparing the microcapsules were based on PEG precipitation from a mixture of molten polymer and essential oil in supercritical CO_2, and PGSS-drying an oil-in-water emulsion of the essential oil with OSA starch. The difference between these processes is the presence of water on the latter, which needs to be removed by carefully tuning the equipment conditions to promote water evaporation. Microcapsules produced by these methods show a mean particle size of 10-500μm for PGSS, and 1-100μm for PGSS-drying. One important observation by scanning electron microscopy (SEM) images is that the experimental conditions can influence the shape of the microparticles. While PEG particles were only spherical (the best shape for controlled release mechanism), in PGSS-drying needle-like structures are formed,, depending on the pre-expansion temperatures of the mixtures, The last one, probably does not hold the active ingredient, presenting some limitations to this specific method without further improvements. Release kinetics were evaluated over a 20-day period. The amount of oil released was proportional to the initial oil concentration on particles, with less than 20% of liberation for low oil concentrations, and about 60% liberation for high oil concentration.

Yang *et al.* [32] assembled polyethylene glycol (PEG) nanoparticles loaded with garlic essential oil using a melt-dispersion method, reaching over 80% of encapsulation efficiency, with round shaped nanoparticles of lower 240nm of average diameter. The encapsulated formulations had their insecticidal activity evaluated against adult red flour beetle *Tribolium castaneum*. While the control experiment done with free garlic oil showed only 11% of efficiency over a five month period, the encapsulated formulation efficiency remained over 80% after five months. This was attributed to the slow and controlled release of the essential oil, and thus could be used as an effective pest control to stored products.

The basic structure of the polymer chitosan was used by Lao *et al.* [33] to build the amphiphilic-modified *N*-(octadecanol-1-glycidyl ether)-*O*-sulfate chitosan (NOSCS). Octadecanol glycidyl ether and sulfate were the hydrophobic and the hydrophilic groups sources respectively. They successfully entrap the herbal insecticide rotenone in the polymer. This chemical compound has been allowed for application in organic crop production due to its natural origin, short persistence in the environment, safety to non-target organisms and low resistance development. The encapsulation was necessary to defeat the problems of chemical stability of the substance to environmental effects and also to improve the solubility of this pesticide in water, which is usually quite low (2.0×10^{-6}g.L^{-1}). Using the reverse micelle method, the authors have assembled nanometric micelles with 167.7-214.0 nm of diameter, with values of critical micellar concentration (CMC) of those chitosan derivatives ranging from 3.55×10^{-3} to 5.50×10^{-3} g.L^{-1}. Although the entrapment efficiency was not very high, they also improved the aqueous solubility of the chemical compound in 13,000 fold, up to 0.026g.L^{-1}, favoring a controlled release of the substance in aqueous media. The complete controlled release took more than 230 hours, almost 10 times more when compared to the chemical compound without nanoencapsulation.

Chitosan derivatives were prepared [34]. They synthesized 6-*O*-carboxymethylated chitosan with anchorage of ricinoleic acid at the *N*-linkage, which further improve its solubility at neutral water (pH = 7.0), to encapsulate the herbal insecticide azadirachtin. Nanoparticles of 200-500nm were obtained by water dispersion with more than 50% of loading efficiency and tested for their stability in outdoor as controlled release systems. Results were compared against simple azadirachtin water dispersion and modified dispersion containing ricinoleic acid and azadirachtin. In 5 days of sun exposure, all content of control samples were lost, while the encapsulated formulation had a nearly constant residual concentration detected throughout the 12 days of the experiment, indicating that the nanoparticles produced were effective at controlling the degradation rate and the release mechanism of the botanical insecticide.

Extracts of Neem were prepared contend high concentration of azadirachtin being nanoencapsulated by Forim *et al.* [35]. Through the use of poly-(ε-caprolactone) polymer, they prepared nanocapsules and nanospheres with average diameter of 150.0 and 250.0nm, respectively. The morphological analysis revealed spherical nanoparticles (Figure 1). The azadirachtin was used as reference. The nanoformulations showed high entrapment efficiency (> 95%) for this compound and a UV stability at least of 30 times more when compared with commercial products.

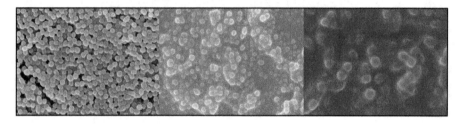

Figure 1. Scanning electron microscopy images of nanoparticles containing extracts of Neem.

2.5. Commercial products

The interesting results obtained in academic researches over the last few decades have been closely followed by several companies. Nevertheless, R&D in nano-based agrochemicals is led mainly by world's largest agroscience companies, further enhancing their market share and consolidating the market structure based on oligopoly that have been seen in late 20th century and early 21st century, when the 10 biggest companies hold around 80% of market [36].

Some companies over the last decade, such as Syngenta, Bayer, Monsanto, Sumitomo, BASF, and Dow Agrosciences have already deposited several different patents comprising a wide range of protocols for production and application of encapsulated formulations, which can be used to produce nanoinsecticides [37-46]. Despite the hard work and heavy investment, no commercial nano-insecticide formulation has been extensively commercialized up to 2012.

Along with those big industries, several other companies, as well as individual researches have been actively depositing patents in the area, thus promoting even more the research and investments in this new field of applied technology. However, as strongly reinforced throughout the world by dozens of organizations such as the ETC Group, the impact of nanotechnology is still unclear, and care should be taken to assure that its use will not bring more problems than solutions [47].

3. Developing new nanopesticides

Many attempts have been made to manage plague insects, for example, using biological control, which is very time consuming. Controlled release systems dawn in this scenario as a very attractive alternative in this battle field.

Controlled release formulations (CRFs) associate the active compound with inert materials. The last ones are responsible for protecting and managing the rate of compound release into the target site in a defined period of time. The main purpose of controlled release systems is ruling the (bio) availability of the active compound after the application [48]. They find the greatest applicabilities in two major agricultural fields: nutrition and protection. In the first one, CRFs are employed in the delivery of fertilizers [49-51]. In the second one, CRFs are mostly used to target plague insects in a sustainable way [52,53], but they can also be applied to block the growth of weeds [54]. Tomioka *et al.*, 2010. Controlled release formulations become especially interesting in cases of antagonist activity of biocides, what can naturally leads to a lower in effectiveness of one or both compounds. In this case the formulation should be "programmed" to release each one at different times [55,56]. Furthermore, still talking about protection, the application of CRFs in wood surfaces, like furniture or floor covering, helps to prevent the deterioration. Van Voris *et al.* [57] patented a formulation in which an insecticide is continually released in a minimum level for a long period of time and is absorbed by the wood. It thus creates a "chemical barrier", blocking the insect attacks.

Most of those controlled release biocides applications were and still are successfully made due to the advances in nanotechnology area.

Micro- and nanomaterials-based formulations are known for some decades. The first microcapsule-based formulation became commercially available in the 1970s [58]. Nanocapsules have been widely used in medicinal area as drug carrier in treatment of diverse diseases [59], from tropical ones [60] up to cancer [61].

Microencapsulation has been used as a versatile tool for hydrophobic pesticides, enhancing their dispersion in aqueous media and allowing a controlled release of the active compound. The use of nanotechnology is a recent approach, and has been a growing subject on several different areas of the science, with an overwhelming perspective. In general, materials that are assembled in nanometric scales (<1000nm) have distinct and almost always better characteristics when compared to the same material built in a conventional manner [62]. One nanometer is a billionth of a meter (1nm = 10^9m). In general, the chemical properties of materials in nanometric scale may be controlled to promote an efficient assemble of a structure which could present several advantages, such as the possibility to better interaction and mode of action at a target site of the plant or in a desired pest due to its tunable controlled release system and larger superficial area, acting as an artificial immune system for plants [34,63]. As smart delivery systems, they confer more selectivity, without hindering in the bioactive compounds towards the target pathogen [65]. Other advantages of the use of nanoparticle insecticides are the possibility of preparing formulations which contain insoluble compounds that can be more readily dispersed in solution. It reduces the problems associated with drifting and leaching, due to its solid nature, and leads to a more effective interaction with the target insect. These features enable the use of smaller amount of active compound per area, as long as the formulation may provide an optimal concentration delivery for the target insecticide for longer times. Since there is no need for re-applications, they also decrease the costs), reduce the irritation of the human mucous-membrane, the phytotoxicity, and the environmental damage to other untargeted organisms and even the crops themselves [65,66]. In a few words, nanotechnology can be applied in several ways in order to enhance efficacy of insecticides in crops.

3.1. Biopolymers

When a commercial formulation for a practical field application is desired, it is very important to employ materials that are compatible with the proposed applications: environment-friendly, readily biodegradable, not generating toxic degradation by-products and low-cost. The use of several biopolymers, i.e., polymers that are produced by natural sources, which at the same time have good physical and chemical properties and still present mild biodegradation conditions, are an interesting approach to avoid the use of petrochemical derivatives that might be another source of environmental contamination. The common polymers (synthetic and natural ones) used in CRFs for insecticides application are listed in Table 1.

3.2. The nanoparticles used in biocides controlled release formulations

The most popular shape of nanomaterials (Figure 2) that have been using in CRFs for biocides delivery are:

a. Nanospheres: aggregate in which the active compound is homogeneously distributed into the polymeric matrix;

b. Nanocapsules: aggregate in which the active compound is concentrated near the center core, lined by the matrix polymer;

c. Nanogels: hydrophilic (generally cross-linked) polymers which can absorb high volumes of water

d. Micelles: aggregate formed in aqueous solutions by molecules containing hydrophilic and hydrophobic moieties.

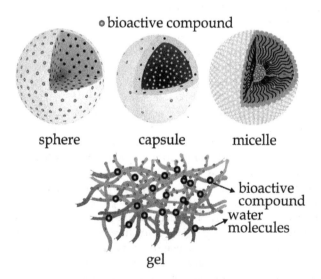

Figure 2. Morphological representation of different nanoparticles.

Polymer	Active compound	Nanomaterial	Ref.
Lignin-polyethylene glycol-ethylcellulose	Imidacloprid	Capsule	[67]
Polyethylene glycol	B-Cyfluthrin	Capsule	[68]
Chitosan	Etofenprox	Capsule	[69]
Polyethylene	Piperonyl Butoxide And Deltamethrin	Capsule	[70]
Polyethylene glycol	Garlic Essential Oil	Capsule	[32]
Poly(acrylic acid)-b-poly(butyl acrylate) Polyvinyl alcohol Polyvinylpyrrolidone	Bifenthrin	Capsule	[71]
Acrylic acid-Bu acrylate	Itraconazole	Capsule	[72]
Carboxymethylcellulose	Carbaryl	Capsule	[73]
Alginate-glutaraldehyde	Neen Seed Oil	Capsule	[74]
Alginate-bentonite	Imidacloprid or Cyromazine	Clay	[75]
Polyamide	Pheromones	Fiber	[76]
Starch-based polyethylene	Endosulfan	Film	[77]
Methyl methacrylate and methacrylic acid with and without 2-hydroxy ethyl methacrylate crosslinkage	Cypermethrin	Gel	[78]
Lignin	Aldicarb	Gel	[79]
Lignin	Imidacloprid Or Cyromazine	Granules	[75]
N-(octadecanol-1-glycidyl ether)-O-sulfate chitosan-octadecanol glycidyl ether	Rotenone	Micelle	[33]
Polyethyleneglycol-dimethyl esters	Carbofuran	Micelle	[80]
Carboxymethyl chitosan-ricinoleic acid	Azadirachtin	Particle[a]	[34]
Chitosan-poly(lactide)	Imidacloprid	Particle[a]	[81]
polyvinylchloride	Chlorpyrifos	Particle[a]	[82]
Cashew gum	*Moringa Oleifera* Extract	Particle[a]	[83]
Chitosan-angico gum	*Lippia Sidoides* Essentioan Oil	Particle[a]	[84]
Polyvinylpyrrolidone	Triclosan	Particle[a]	[85]
Anionic surfactants (sodium linear alkyl benzene sulfonate, naphthalene sulfonate condensate sodium salt and sodium dodecyl sulfate)	Novaluron	Powder	[86]
Vinylethylene and vinylacetate	Pheromones	Resin	[87]
Glyceryl ester of fatty acids	Carbaryl	Spheres	[15]
Poly(ε-caprolactone)	Active Ingredients[b]	Spheres	[88]
Poly(methyl methacrylate)-poly(ethylene glycol) Polyvinylpyrrolidone	Carbofuran	Suspension	[89]

[a] The authors do not mention which active compounds they encapsulated in the nanospheres; [b] The authors do not mention if the particles are spheres or capsules

Table 1. Several examples of polymers often used in the nanoparticle production.

Dendrimers, nanoclays, nanopowders and nanofibers are other possible formulations which might be used during nano or microparticle production[75, 76, 86, 90]. On the other hand, nanotubes are mostly applied in plants improvement. The polymeric nanoparticles and gels are by far the mostly used for insecticides application, because they have an extra advantage of being biodegradable.

3.3. Methods for preparation of nanomaterials based controlled-release formulations for biocides application

According to Wilkins [48], the methods for CRF preparation can be separated in chemical or physical ones (Figures 3 and 4, respectively).

The chemical methods are based on a chemical bond (usually a covalent one) formed between the active compound and the coating matrix, such as a polymer. This bound can be placed in two different sites: in the main polymeric chain or in a side chain. In the first one, the new "macromolecule" is also called a pro-biocide, because the compound will get its properties in fact when it is released. In the second one, the insecticide molecule can bind initially to the side-chain of one monomer and then the polymerization reaction takes place or the polymerization occurs first and only after that, the biocide binds to the side chain. There is still a third way, based on the intermolecular interactions. In this case, the biocide is "immobilized" in the net produced by the cross-linkages in the polymer.

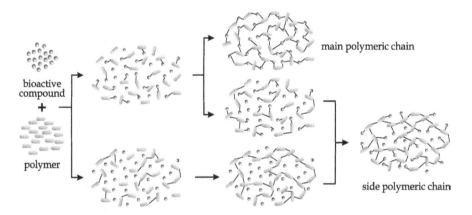

Figure 3. Chemical methods for CRF preparation

The physical methods can also be split in two distinct categories. In the first, a mixture of biocide and polymer is made. As the last has a higher energy density, it moves to a more external layer, forming a kind of monolithic structure. In the other one, the polymeric chain forms a "membrane" isolating the bioactive compound from the external environment. This is the method which will produce the nanocapsules themselves.

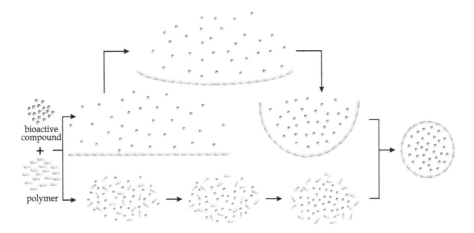

Figure 4. Physical methods for CRF preparation

Although there are some different kinds of nanomaterials that can be used in CR formulations, the micro- and nanocapsules are by far the most widely used for controlled release of biocides. For this reason, the techniques described here will be restricted to micro and nano-encapsulation process.

3.4. Micro and nanoencapsulation techniques

The first formulation containing polymeric-based nanocarriers for controlled release of biocides dates from the early 1970's [11,92]. Recently, John *et al.* [93] reviewed the most commonly techniques used to prepare micro- and nanocapsules containing microorganisms (for this kind of application, see section 2.3). However, the techniques they commented can be also utilized to prepare nanocapsules for insecticides application in general. Shahidi and Han [94] and Wilkins [48] classified them as physicochemical, chemical or physical process-based. Some are described below.

3.4.1. The physicochemical-based techniques

a. *Emulsion*: This technique is used to produce a system of two immiscible liquid phases (water and oil), where one (the dispersed phase) is dispersed into the other (continuous phase) in a controlled way (usually in a dropwise one). The bioactive compound (usually water-soluble) and the polymer are solubilized each one in a phase (water or oil). One of the solutions is gradually dripped into the other under vigorous stirring. After the homogenization, the emulsion is formed. If the oil is the dispersed phase, the emulsion is classified as O/W (oil/water). If it is water, the emulsion is called W/O (water/oil) [95]. The emulsion itself also represents a crucial step for some other more complexes preparation ones.

b. *Coacervation*: This process is based on the reduction of polymer's solubility. According to Wilkins[48] the encapsulation goes through a separation of phases and can be simple or complex. In a simple coacervation, the addition of an external agent, like a salt or wa-ter-miscible solvent, to an aqueous solution containing a hydrophilic polymer-insecti-cide complex causes its precipitation. Complex coacervation involves opposite charges and electrostatic attraction. A solution containing different ionizable polymers is sub-mitted to a pH change. The polymers turn positively or negatively charged. The electro-static attractive forces between the opposite charges become much stronger than the particle-solvent intermolecular ones, leading to the copolymer precipitation.

c. *Emulsion-solvent evaporation*: According to Iwata and McGinity [96] this technique com-prises two or three steps. In the first one, an O/W (or W/O) emulsification must be ini-tially formed. The polymer is usually solubilized in the dispersed phase. If the emulsion has only two components like this one, it is called a single emulsion. For this type, the whole process has only two steps and the first one ends here. However, there is also other type, called double emulsion, represented as W/O/W', where the emulsion al-ready prepared in the first step is dispersed into an organic solvent, like acetonitrile. In this case, the aqueous solution containing the active compound is dripped in an oil phase (usually a vegetable oil), under stirring. This emulsion is then dispersed, under stirring, in an organic solvent solution containing the polymer. The last step, common for single and double emulsion, is the evaporation of the solvent, what can be per-formed at room temperature or under reduced pressure. After solvent removal, the par-ticles are ready for use.

d. *Emulsion crystallization/ solidification*: According to the procedure published by Iqbal *et al.* [97], an emulsion is initially prepared as already described in this section. The only difference remains in the temperature in which it is made. The authors prepared the emulsion at 60°C. The next step is crucial for technique success. The warm emulsion is pumped through a capillary partially immersed in a coolant liquid (temperature: 10°C). At the capillary exit, the emulsion forms spherical drips which move to raise the cooling liquid's surface. The drop is cooled down during the course, solidifying and forming the particles which are collected at the top.

e. *Diffusion-controlled emulsion*: In this process, a monomer rich phase is laid over the aque-ous solution containing the insecticide, under a smooth stir. The monomers then diffuse into the aqueous fase, "trapping" the bioactive molecules in a micellar structure [98].

f. *Liposome entrapment*: Some protocols to prepare liposomes are described by Mozafari *et al.* [99]. The standard one is resumed here.

In the first step, an organic solution (chloroform or methanol ones) containing hydrophobic molecules such phospholipids and cholesterol is prepared. The solvent then is evaporated forming a thin film. Next, an aqueous solution containing the bioactive compound is spread over this film. Some mechanical or thermal perturbation like ultrasound or heating is ap-plied to the system to promote the formation of single or double layer sheet. The sheet will detach from the support, closing itself, forming the liposomes. During this closing process, the sheet traps the biocides molecules.

3.4.2. The chemical techniques

a. *Interfacial polymerization*: As the name says, this technique is based in a polymerization reaction which occurs in an interface of two immiscible liquids. According to Wilkins [48], polymerization can occur through an addition or condensation reaction. In the mostly addition-governed process, the polymerization starts in the oil phase, where the monomers and insecticide are dispersed. However, the reaction only takes place when it is catalyzed by free radicals, which are dissolved in the aqueous phase. In condensation-governed process (the most suitable route for biocides nanoencapsulation), the reactive monomers are dissolved each one in a different phase. As the dispersed phase is dripped into the continuous phase, the reaction occurs in the droplet interface, producing the polymer. When a solvent with a low boiling temperature is used as the oil phase (either in dispersed or continuous one) and contains the monomers dissolved, the process is a little different. After the dripping, the system is heated. The solvent thus evaporates, leaving the particles that, due to the water insolubility, precipitates. This particular technique variation can also be called interfacial polymer deposition [100].

b. *Molecular inclusion*: This technique is used to increase the solubility of water-insoluble compounds in aqueous solution. Macromolecules like cyclodextrins [101] have an inner hydrophobic face and an outer hydrophilic face. An oil phase containing the biocide is dripped, under continuous stirring, into the aqueous macromolecule solution. During the dripping, the macromolecule "traps" the insecticide molecules via intermolecular interactions.

3.4.3. The physical techniques

a. *Extrusion*: The bioactive compound is mixed with hydrocolloids and then, the colloid is squeezed out under pressure. The pressure during the process should be adjusted according to the viscosity of colloids.

b. *Spray drying*: This technique is based in solvent evaporation at high temperatures. The spray drying process has already been described in details by Ré [102]. The following text is only a brief resume. Initially, the active compound and the polymeric matrix are solubilized in their respective solvents, which should not be miscible. Then, they are mixed under vigorous stirring to form an emulsion (or dispersion whether one of the components is in the solid state). The emulsion undergoes an atomization to produce droplets. In the next step, the droplets are submitted to a hot air flow that forces the solvent (generally water) evaporation, leaving only a dry powder. The greatest advantage of this technique is that it can be easily scaled up for a large scale nanocapsules production.

c. *Freeze drying*: This technique is also known as liophilization. It is the opposite of the spray drying, because it uses a low temperature system. A suspension or emulsion is prepared to enable the polymer-insecticide formation. For emulsions, an additional step is required before the execution of the technique: the removal of the oil or organic solvent under reduced pressure. For both (emulsion and suspension), the aqueous phase is

frozen and submitted to a low pressure system. When the pressure is drastically reduced, the water sublimes (goes from solid to vapor state), leaving only the particles.

3.5. Mechanism of biocide release

In the paper published by Kratz et al. [103] the text begins with the statement: "Nanoparticles only start working after they are placed in a desired location". In other words, an efficient CR formulation must remain inactive until the active compound is released.

The way how an inert material, such the nanopolymers, controlss the amount and rate a
chemical is released is object of study since the late 1960's [104] and early 1970's [105].

How the release of the bioactive compound occurs depends basically on the chemical nature
of the formulation. In various polymeric nanomaterials, the controlled release proceeds via
diffusion. It does not matter if the bioactive compound is dissolved (micro- or nanospheres)
or if it is encapsulated (micro or nanocapsules). The process does not depend on the chemical structure of the formulation constituents [11] neither on the intermolecular interactions.
The rate control is made based on the interactions between the carrier and the biocide. The
stronger the interaction will be slower the release rate. In the 1990's, the release dynamics
was investigated via the use of ^{14}C-labelled molecules of herbicides [106,107]. Qi et al. [107]
studied the dynamic of controlled release for herbicides. They used ^{14}C-labelled molecules of
benthiocarb and butachlor and observed that the release is made by a diffusive process.
Some years later and without any radiolabeled molecules, Fernandez-Perez et al. [108]
found the same results. They prepared a granule-based CRF constituted by lignin and imidacloprid. They measured the amount of compound released in water under a dynamic
flow condition during a defined period of time. The data fitted a diffusion curve based on
the model proposed by Ritger and Papas [109,110]. Since then, other similar studies have
been published [111-114].

Some other polymeric nanomatrixes, especially those formed by a carboxylic acid and a met-
allic cation, can be disassembled when in contact with water, releasing the bioactive compound [92]. The release rates depend on the physicochemical characteristic of both
molecules. The more hydrophobic the polymer slower will be the bioactive compound release. The same applies to the last one: the higher water-solubility, faster it will be released.
The formulation itself also affects directly the release rate. In water-based one, the rate control tends to disappear, due to the matrix (or support) degradation. If the particles are solubilized in an organic solvent, like acetone, the formulation becomes sticky and the release
rate slows down. A granule-based formulation sounds more efficient. It can be applied direct to the soil and the bioactive compound will be released according to the soil moisturize
(water content), leading to a long lasting control.

In other formulations, the bioactive compound is covalently bound to the polymeric matrix
[115]. To the release takes place, a chemical interaction must be broken. It usually occurs via a
hydrolysis reaction, what affects many polymer-insecticide bounds in a chain reaction. The release control depends on the strength of those chemical bounds, the chemical properties of
both molecules and on the size and structure of the macromolecule formed [11]. The higher the

biocompound solubility in water, faster the reaction occurs. Concerning the chemical properties of the polymer, Allan *et al.* [11] studied the differences in the release kinetics when 2-methyl-4-chlorophenoxy acetic is chemically bound to polyvinylalcohol (a water-soluble polymer) or when it is bound to cellulose or lignin (water-insoluble polymers). In the first situation, the level of the applied herbicide tends to go down, because the equilibrium

Polymer – insecticide ⇌ *Polymer + Insecticide*

will always exist. I the last situation, as the "free polymer" is water–insoluble, the equilibrium moves towards the right side and the level of the applied herbicide tends to go up (Figure 5).

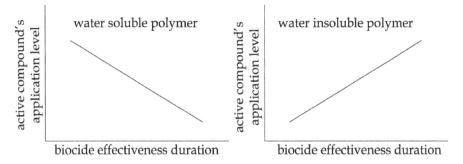

Figure 5. Trend in active compound's application rate (*Adapted* from [11]).

Whatever the mode of liberation, it should be kept in mind that controlled release formulations have a limited maximum amount for the release of the biocide[116]. This means that the total amount of released product may not be necessarily equal to the amount of chemicals incorporated to the formulation neither to the amount of free applied product[117]. This is the reason why the concentration of the active compound in a CRF is usually higher than in a conventional one. However it does not contradict what was said earlier about advantageous reduced amount if biocide applied, since the number of applications should be smaller.

Studies recently published suggested that the encapsulation of biocides reduces their toxicity [117,118]. However, many issues regarding the toxicity of the nanomaterial themselves towards the environmental and even the worker's health remains unclear [119].

4. Conclusion

The increasing worldwide demand for foods requires modern techniques of agricultural production minimizing losses in the crops, transportation and storage. Among the main causes of agricultural losses there are the plague insects. Insecticides are an important control tool. However, some collateral effects may be credited to their indiscriminate use such as environmental contamination, human poisoning, reduction in the number of natural enemies, insecticide resistance by plague insects, etc.

In this scenario, nano- and microparticles have been reaching a prominent position. Formulations containing insecticides have been prepared in colloidal suspensions or powder, in nano or micro scale, where they present several advantages such as increasing stability of the active organic compound (UV, thermal, hydrolysis, etc.), foliar settling, reduction in foliar leaching, systemic action, synergism, specificity, etc. As consequence, the amount of insecticide necessary (dosage), the number of applications, human exposure to insecticides and environmental impact are reduced. The nano- and microformulations have been employed not only for synthetic insecticides but also in alternative products to control plague insects such as natural products (herbal extracts) and entomopathogenic microorganisms.

In order to prepare nano- and microformulations, several chemical and physical techniques have been developed. In general, they should be prepared by using polymeric materials which are biocompatible and biodegradable. This practice has the aim to avoid the emergence of new environmental and toxicological problems. The biopolymers are produced by microorganisms, synthesis or even petroleum derivate products. In common, when exposed to the environment they are easily destroyed by UV radiation and/or microorganism enzymes generating CO_2 and H_2O as final product. The degradation processes of biopolymers may lead, or not, to the release mechanisms of active organic compounds of a nano- or microparticles. Processes such as swelling, hydrolysis, diffusion, erosion, etc., must be manipulated in a controlled way in order to obtain the desired characteristics of application and biological activity for the formulated products.

As a result of the application of these new nano- and micro- technologies, which have been quickly developed due to new sensitive analytical technologies of characterization, new ways to control plague insects are emerging, thinking not only in lethal action on the target insect, but also in all ecosystems, which include fishes, natural enemies, vegetation, microorganisms, animals, the man himself, etc.

Acknowledgments

The authors are grateful to Conselho Nacional de Desenvolvimento Científico e Tecnológico (CNPq) and to Fundação de Amparo à Pesquisa do Estado de São Paulo (FAPESP), for the financial support.

Author details

Bruno Perlatti, Patrícia Luísa de Souza Bergo, Maria Fátima das Graças Fernandes da Silva, João Batista Fernandes and Moacir Rossi Forim

Federal University of São Carlos, Department of Chemistry, Brazil

References

[1] Chapman R.F. Foraging and Food Choice in Phytophagous Insects. In: Hardege J.D. (ed.) Chemical Ecology. Oxford: Eolss Publishers; 2009. Available from http://www.eolss.net/Sample-Chapters/C06/E6-52-02-02.pdf (Acessed 18 august 2012).

[2] Ware GW, Whitacre DM. The Pesticide Book, 6th ed. Ney York: Meister Publishing Company; 2004.

[3] DeVilliers SM, Hoisington DA. The Trends and Future of Biotechnology Crops for Insect Pest Control. African Journal of Biotechnology 2011;10(23) 4677-4681.

[4] Panagiotakopulu E, Buckland PC, Day PM, Sarpaki, AA, Doumas C. Natural Insecticides and Insect Repellents in Antiquity: A review of the Evidence. Journal of Archaeological Science, 1995;22 705-710.

[5] Ghormade V, Deshpande MV, Paknikar KM. Perspectives for Nano-Biotechnology Enabled Protection and Nutrition of Plants. Biotechnology Advances 2011;29 792-903.

[6] Mogul MG, Akin H, Hasirci N, Trantolo DJ, Gresser JD, Wise DL. Controlled Release of Biologically Active Agents for Purposes of Agricultural Crop Management. Resources, Conservation And Recycling 1996;16 289-320.

[7] Keifer MC. Effectiveness of Interventions in Reducing Pesticide Overexposure and Poisonings. American Journal of Preventive Medicine 2000;18(4) 80–89.

[8] Akelah A. Novel Utilizations of Conventional Agrochemicals by Controlled Release Formulations. Materials Science and Engineering C 1996;4 83-98.

[9] Casanova H, Araque P, Ortiz C. Nicotine Carboxylate Insecticide Emulsions: Effect of the Fatty Acid Chain Length. Journal of Agricultural and Food Chemistry 2005;53 9949-9953.

[10] Wang L, Li X, Zhang G, Dong J, Eastoe J. Oil-in-Water Nanoemulsions for Pesticide Formulations. Journal of Colloid and Interface Science, 2007;314 230–235.

[11] Allan GG, Chopra CS, Neogi AN, Wilkins RM. Design and Synthesis of Controlled Release Pesticide-Polymer Combinations. Nature 1971;234 349-351.

[12] Greene LC, Meyers PA, Springer JT, Banks PA. Biological Evaluation of Pesticides Released from Temperature-Responsive Microcapsules. Journal of Agricultural and Food Chemistry 1992;40 2274-2270.

[13] Latheef MA, Dailey Jr OD, Franz E. Efficacy of Polymeric Controlled Release Formulations of Sulprofos Against Tobacco Budworm, Heliothis virescens (Lepidopetea:Noctuidae) on Cotton. In: Berger PD, Devisetty BN, Hall FR. (eds.) Pesticide Formulations and Applications Systems: 13th volume, ASTM STP 1183. Philadelphia: American Society for Testing and Materials; 1993. p. 300-311.

[14] Arthur FH. Evaluation of an Encapsulated Formulation of Cyfluthrin to Control Sitophilus oryzae (L.) on Stored Wheat. Journal of Stored Products Research 1999;35 159-166.

[15] Quaglia F, Barbato F, De Rosa G, Granata E, Miro A, La Rotonda MI. Reduction of the Environmental Impact of Pesticides: Waxy Microspheres Encapsulating the Insecticide Carbaryl. Journal of Agricultural and Food Chemistry 2001;49 4808-4812.

[16] Cao Y, Huang L, Chen J, Liang J, Long S, Lu Y. Development of a Controlled Release Formulation Based on a Starch Matrix System. International Journal of Pharmaceutics 2005;298 108–116.

[17] Takei T, Yoshida M, Hatate Y, Shiomori K, Kiyoyama S. Preparation of Polylactide/ Poly(Є-Caprolactone) Microspheres Enclosing Acetamiprid and Evaluation of Release Behavior. Polymer Bulletin 2008;61 391–397.

[18] Tsuda N, Ohtsubo T, Fuji M. Preparation of Self-Bursting Microcapsules by Interfacial Polymerization. Advanced Powder Technology 2011; doi:10.1016/j.apt. 2011.09.005.

[19] Tsuda N, Ohtsubo T, Fuji M, Study on the Breaking Behavior of Self-Bursting Microcapsules Advanced Powder Technology 2012; doi:10.1016/j.apt.2011.11.006.

[20] Guan H, Chi D, Yu J, Li H. A novel photodegradable insecticide: Preparation, Characterization and Properties Evaluation of Nano-Imidacloprid. Pesticide Biochemistry and Physiology 2008;92 83–91.

[21] Guan H, Chi D, Yu J, Li H. Dynamics of Residues From a Novel Nano-Imidacloprid Formulation in Soybean Fields. Crop Protection 2010;29 942-946.

[22] Ramírez-Lepe M, Aguilar O, Ramírez-Suero M, Escudero B. Protection of the Spore-Toxin Complex of bacillus thurigiensis serovar israelensis from Ultraviolet Irradiation with Aluminum-cmc Encapsulation and Photoprotectors. Southwestern Entomologist 2003;28(2) 137-143.

[23] Tamez-Guerra P, McGuire MR, Behle RW, Shasha BS, Galán-Wong LJ. Assessment of Microencapsulated Formulations for Improved Residual Activity of Bacillus thuringiensis. Journal of Economic Entomology 2000;93 219–225.

[24] Tamez-Guerra P, McGuire MR, Behle RW, Hamm JJ, Sumner HR, Shasha, B.S. Sunlight Persistence and Rainfastness of Spray-Dried Formulations of Baculovirus Isolated from Anagrapha falcifera (Lepidoptera: Noctuidae). Journal of Economic Entomology 2000;93 210–218.

[25] Tamez-Guerra P, McGuire MR, Behle RW, Shasha BS, Pingell RL. Storage stability of Anagrapha falcifera nucleopolyhedrovirus in spray-dried formulations. Journal of Invertebrate Pathology 79 2002 7-16.

[26] Behle RW, Tamez-Guerra P, McGuire MR. Evaluating Conditions for Producing Spray-Dried Formulations of Anagrapha falcifera Nucleopolyhedroviruses (AfMNPV). Biocontrol Science and Technology 2006;16 941–952.

[27] McGuire MR, Tamez-Guerra P, Behle RW, Streett DA. Comparative Field Stability of Selected Entomopathogenic Virus Formulations. Journal of Economic Entomology 2001;94(5) 1037-1044.

[28] Arthurs SP, Lacey LA, Behle RW. Evaluation of Spray-Dried Lignin-Based Formulations and Adjuvants as Solar Protectants for the Granulovirus of the Codling Moth, Cydia pomonella (L). Journal of Invertebrate. Pathology 2006;93 88–95.

[29] [29]Behle RW, Popham HJR. Laboratory and Field Evaluations of the Efficacy of a Fast-Killing Baculovirus Isolate From Spodoptera frugiperda. Journal of Invertebrate Pathology 2012;109 194–200.

[30] Varona S, Martín Á, Cocero MJ. Formulation of a Natural Biocide Based on Lavandin Essential Oil by Emulsification Using Modified Starches. Chemical Engineering Process 2009;48 1121–1128.

[31] Varona S, Kareth S, Martín Á, Cocero MJ. Formulation of Lavandin Essential Oil with Biopolymers by PGSS for Application as Biocide in Ecological Agriculture. Journal of Supercritical Fluids 2010;54 369–377.

[32] Yang FL, Li XG, Zhu F, Lei CL. Structural Characterization of Nanoparticles Loaded with Garlic Essential Oil and Their Insecticidal Activity Against Tribolium castaneum (Herbst) (Coleoptera: Tenebrionidae). Journal of Agricultural and Food Chemistry 2009;57(21) 10156-10162.

[33] Lao SB, Zhang ZX, Xu HH, Jiang GB. Novel Amphiphilic Chitosan Derivatives: Synthesis, Characterization and Micellar Solubilization of Rotenone. Carbohydrate Polymers 2010;82 1136–1142.

[34] Feng BH, Peng LF. Synthesis and Characterization of Carboxymethyl Chitosan Carrying Ricinoleic Functions as an Emulsifier for Azadirachtin. Carbohydrate Polymers 2012;88 576– 582.

[35] Forim M.R., da Silva M.F.G.F, Fernandes J.B. Secondary Metabolism as a Measure of Efficacy of Botanical Extracts: The use of Azadirachta indica (Neem) as a Model. In: Perveen F. (ed.) Insecticides - Advances in Integrated Pest Management. Rijeka: InTech; 2011. p367-390. Available from

[36] ObservatoryNANO. Nanotechnologies for Nutrient and Biocide Delivery in Agricultural Production. Working Paper Version. 2010. Available from

[37] Feistel L, Halle O, Mitschker A, Podszun W, Schmid C. inventors; Bayer AG., assignee. Production of monodisperse crosslinked polystyrene beads, useful for producing ion exchangers, comprises a seed-feed method including a micro-encapsulation step to reduce soluble polymer content. US Patent Number 6365683-B2. 2002 Apr 2.

[38] van Koppenhagen JE, Scher HB, Lee KS, Shirley IM, Wade P, Follows R., inventors; Syngenta Limited., assignee. Acid-triggered microcapsules. US Patent Number 6514439-B2. 2003a Fev 4.

[39] van Koppenhagen JE, Scher HB, Lee KS, Shirley IM, Wade P, Follows R., inventors; Syngenta AG Limited., assignee. Base-triggered release microcapsules. US Patent Number 6544540-B2. 2003b Apr 8.

[40] Christensen B, Suty-Heinze A, Schick N, Wolf H. inventors; Bayer AG., assignee. Microcapsule suspension for agrochemical use comprises microcapsule envelope formed from tolylene diisocyanate and 4,4'-methylene-bis-(cyclohexyl-isocyanate) mixtures by reaction with diamine or polyamine. US Patent Number 6730635-B2. 2004 May 4.

[41] Keiichiro I, Tahahiro S., inventors; Sumitomo Chemical., assignee. Micro-encapsulated insecticide composition, especially useful for controlling termites or cockroaches, comprising pyrethrin encapsulated in synthetic polymer, preferably polyurethane or polyuria. Patent Number FR2846853. 2004 May 14.

[42] Botts MF, Kohn FC, Miller ML. inventors; Monsanto Company, assignee. Particles containing agricultural active ingredients. US Patent Number 7070795-B1. 2006, Jul 4.

[43] Schrof W, inventor; BASF AG., assignee. Nanoparticles comprising a crop protection agent. Patent Number EP1465485-B1. 2006 Apr 26.

[44] Wilson SL, Boucher Jr. RE., inventors; Dow Agrosciences LLC., assignee. Pesticide formulation, useful for controlling an insect population such as aphids and beet army worm, comprises an organophosphate pesticide e.g. diazinon, dimethoate and disulfoton, and a polymer forming a capsule wall. US Patent Number 2011/0052654 A1. 2010 Mar 1.

[45] Graham MC, King JE, Logan MC, Wujek DG., inventors; Dow AgroSciences LLC., assignee. Pesticide Compositions. US Patent Number 8021675. 2011 Set 20.

[46] Slater R, Alfred R, Peter M, Phillippe C, Gaume A., inventors; Syngenta AG., assignee. Use of spiroheterocyclic compounds for controlling insects from order hemiptera and that are resistant to neonicotinoid insecticides e.g. Myzus persicae, and for protecting useful plants such as cereals or fruit plants. WO/20111/51249. 2011 Dec 8.

[47] ETC Group. Down on the Farm: The Impact of Nano-Scale Technologies on Food and Agriculture. Winnipeg: ETC Group; 2004.

[48] Wilkins RM. Controlled Release Technology, Agricultural. In: Seidel A. (ed.) Kirk-Othmer Encyclopedia of Chemical Technology 5th Ed. New Jersey: John Wiley & Sons; 2004.

[49] Xie L, Liu M, Ni B, Zhang X, Wang Y. Slow-Release Nitrogen and Boron Fertilizer from a Functional Superabsorbent Formulation Based on Wheat Straw and Attapulgite. Chemical Engineering Journal 2011;167 342-348.

[50] Xie L, Liu M, Ni B, Wang Y. Utilization of Wheat Straw for the Preparation of Coated Controlled-Release Fertilizer with the Function of Water Retention. Journal of Agricultural and Food Chemistry 2012;60 6921-6928.

[51] Wang Y, Liu M, Ni B, Xie L. κ-Carrageenan-Sodium Alginate Beads and Superabsorbent Coated Nitrogen Fertilizer with Slow-Release, Water-Retention, and Anti-compaction Properties. Industrial & Engineering Chemistry Research 2012;51 1413-1422.

[52] Park M, Lee CI, Seo YJ, Woo SR, Shin D, Choi J. Hybridization of the Natural Antibiotic, Cinnamic Acid, with Layered Double Hydroxides (LDH) as Green Pesticide. Environmental Science And Pollution Research 2010;17 203-209.

[53] Fujii T, Hojo T, Ishibashi N, Saguchi R, Fukumoto T. Sustained release pheromone preparation, having carboxylic acid as pheromone substance, for targeting insect pest. Patent number US 20120156165 A1 20120621. 2012.

[54] Tomioka A, Sugiyama M, Suda Y, Kadokura K. Improved release agrochemical formulation of herbicide, aluminum salt and silicate mineral. Patent number WO 2010003499 A2 20100114. 2010.

[55] Frisch G, Bickers U, Young KA, Hacker E, Schnabel G. Sustained-release combinations of herbicides with anionic polymers. Patent number WO 2001084926 A1 20011115. 2001.

[56] Krause HP, Schnabel G, Frisch G, Wuertz J, Bickers U, Hacker E, Auler T, Melendez A, Haase D. Sustained-release combinations of carrier-incorporated pesticides. Patent number WO 2001084928 A1 20011115. 2001.

[57] van Voris P, Cataldo DA, Burton FG. Controlled-Release Insecticidal Wood Preservative. Patent number US 6852328 B1 20050208. 2005.

[58] Fanger G.O. Microencapsulation: A Brief History and Introduction. In: Vandegaer J.E. (ed.) Microencapsulation: Process and Applications. Plenum Press: New York; 1974. p1-20.

[59] Radhika PR, Sasikanth and Sivakumar T. Nanocapsules: a New Approach in Drug Delivery. International Journal of Pharmaceutical Sciences and Research 2011;2 1426-1429.

[60] Kuntworbe N, Martini N, Shaw J, Al-Kassas R. Malaria Intervention Policies and Pharmaceutical Nanotechnology as a Potential Tool for Malaria Management. Drug Development Research 2012;73 167-184.

[61] Joshi MD, Unger WJ, Storm G, van Kooyk Y, Mastrobattista E. Targeting Tumor Antigens to Dendritic Cells Using Particulate Carriers. Journal of Controlled Release 2012;161, 25-37.

[62] Moraru C, Panchapakesan C, Huang Q, Takhistov P, Liu S, Kokini J. Nanotechnology: A New Frontier in Food Science. Food Technology 2003;57(12) 24-29.

[63] Pérez-de-Luque A, Rubiales D. Nanotechnology for Parasitic Plant Control. Pest Management Science 2009;65 540–545.

[64] Brausch KA, Anderson TA, Smith PN, Maul JD. Effects of Functionalized Fullerenes on Bifenthrin and Tribufos Toxicity to Daphnia magna: Survival, Reproduction, and Growth Rate. Environmental Toxicology and Chemistry 2010;29 2600-2606.

[65] Peteu SF, Oancea F, Sicuia OA, Constantinescu F, Dinu S. Responsive Polymers for Crop Protection. Polymers, 2010;2 229-251.

[66] Margulis-Goshen K, Magdassi S. Nanotechnology: An Advanced Approach to the Development of Potent Insecticides. In: Ishaaya I, Horowit,z AR, Palli SR. (eds.) Advanced Technologies for Managing Insect Pests. Dordrecht: Springer; 2012. p. 295-314.

[67] Flores-Cespedes F, Figueredo-Flores CI, Daza-Fernandez I, Vidal-Pena F, Villafranca-Sanchez M, Fernandez-Perez M. Preparation and Characterization of Imidacloprid Lignin-Polyethylene Glycol Matrices Coated with Ethylcellulose. Journal of Agricultural and Food Chemistry 2012;60 1042-1051.

[68] Loha KM, Shakil NA, Kumar J, Singh MK, Srivastava C. Bio-efficacy Evaluation of Nanoformulations of β-cyfluthrin Against Callosobruchus maculatus (Coleoptera: Bruchidae). Journal of Environmental Science and Health Part B-Pesticides Food Contaminants and Agricultural Wastes 2012;47 687-691.

[69] Hwang IC, Kim TH, Bang SH, Kim KS, Kwon HR, Seo MJ, Youn YN, Park HJ, Yasunaga-Aoki C, Yu YM. Insecticidal Effect of Controlled Release Formulations of Etofenprox Based on Nano-Bio Technique. Journal of the Faculty of Agriculture Kyushu University 2011;56 33-40.

[70] Frandsen MV, Pedersen MS, Zellweger M, Gouin S, Roorda SD, Phan TQC. Piperonyl butoxide and deltamethrin containing insecticidal polymer matrix comprising HDPE and LDPE. Patent number WO 2010015256 A2 20100211. 2010.

[71] Liu Y, Tong Z, Prud'homme RK. Stabilized Polymeric Nanoparticles for Controlled and Efficient Release of Bifenthrin. Pest Management Science 2008;64 808-812.

[72] Goldshtein R, Jaffe I, Tulbovich B. Hydrophilic dispersions of nanoparticles of inclusion complexes of amorphous compounds. Patent number US 20050249786 A1 20051110. 2005.

[73] Isiklan N. Controlled Release of Insecticide Carbaryl from Crosslinked Carboxymethylcellulose Beads. Fresenius Environmental Bulletin 2004;13 537-544.

[74] Kulkarni AR, Soppimath KS, Aminabhavi TM, Dave AM, Mehta MH. Application of Sodium Alginate Beads Crosslinked with Glutaraldehyde for Controlled Release of Pesticide. Polymers News 1999;2 285-286.

[75] Fernandez-Perez M, Garrido-Herrera FJ, Gonzalez-Pradas E. Alginate and Lignin-Based Formulations to Control Pesticides Leaching in a Calcareous Soil. Journal of Hazardous Materials 2011;190 794-801.

[76] Hellmann C, Greiner A, Wendorff JH. Design of Pheromone Releasing Nanofibers for Plant Protection. Polymers for Advanced Technologies 2011;22 407-413.

[77] Jana T, Roy BC, Maiti S. Biodegradable Film. 6. Modification of the Film for Control Release of Insecticides. Euroreian Polymer Journal 2001;37 861-864.

[78] Rudzinski WE, Chipuk T, Dave AM, Kumbar SG, Aminabhavi TM. pH-sensitive Acrylic-Based Copolymeric Hydrogels for the Controlled Release of a Pesticide and a Micronutrient. Journal of Applied Polymer Science 2003;87 394-403.

[79] Kok FN, Wilkins RM, Cain RB, Arica MY, Alaeddinoglu G, Hasirci V. Controlled Release of Aldicarb from Lignin Loaded Ionotropic Hydrogel Microspheres. Journal of Microencapsulation 1999;16 613-623.

[80] Shakil NA, Singh MK, Pandey A, Kumar J, Parmar VS, Singh MK, Pandey RP, Watterson AC. Development of Poly(Ethylene Glycol) Based Amphiphilic Copolymers for Controlled Release Delivery of Carbofuran. Journal of Macromolecular Science, Part A: Pure and Applied Chemistry 2010;47 241-247.

[81] Li M, Huang Q, Wu Y. A novel Chitosan-Poly(Lactide) Copolymer and Its Submicron Particles as Imidacloprid Carriers. Pest Management Science 2011;67 831-836.

[82] Liu Y, Laks P, Heiden P. Controlled Release of Biocides in Solid Wood. II. Efficacy Against Trametes versicolor and Gloeophyllum trabeum Wood Decay Fungi. Journal of Applied Polymer Science 2002;86 608-614.

[83] Paula HCB, Rodrigues MLL, Ribeiro WLC, Stadler AS, Paula RCM, Abreu FOMS. Protective Effect of Cashew Gum Nanoparticles on Natural Larvicide from Moringa oleifera Seeds. Journal of Applied Polymer Science 2012;124 1778-1784.

[84] Paula HCB, Sombra FM, Abreu FOMS, de Paula, RCM. Lippia sidoides Essential Oil Encapsulation by Angico Gum/Chitosan Nanoparticles. Journal of the Brazilian Chemical Society 2010;21 2359-2366.

[85] [85]Narayanan KS, Jon D, Patel J, Winkowski K. Aqueous composition for delivering bioactive hydrophobic nanoparticles. Patent number WO 2008016837 A2 20080207. 2008.

[86] Elek N, Hoffman R, Raviv U, Resh R, Ishaaya I, Magdassi S. Novaluron Nanoparticles: Formation and Potential Use in Controlling Agricultural Insect Pests. Colloids and Surfaces A-Physicochemical and Engineering Aspects 2010;372 66-72.

[87] Wright JE. Formulation for insect sex pheromone dispersion. Patent number US 5670145 A 19970923. 1997.

[88] Le Roy Boehm AL, Zerrouk R, Fessi H. Poly-ε-caprolactone Nanoparticles Containing a Poorly Soluble Pesticide: Formulation and Stability Study. Journal of Microencapsulation 2000;17 195-205.

[89] Chin CP, Wu HS, Wang SS. New Approach to Pesticide Delivery Using Nanosuspensions: Research and Applications. Industrial & Engineering Chemistry Research 2011;50 7637-7643.

[90] Hayes RT, Owen JD, Chauhan AS, Pulgam VR. PEHAM dendrimers for use in agricultural formulations. Patent number WO 2011053605 A1 20110505. 2011.

[91] Srinivasan C, Saraswathi R. Nano-Agriculture - Carbon Nanotubes Enhance Tomato Seed Germination and Plant Growth. Current Science India 2010;99 274-275.

[92] Beasley ML, Collins RL. Water-Degradable Polymers for Controlled Release of Herbicides and Other Agents. Science 1970;169 769-770.

[93] John RP, Tyagi RD, Brar SK, Surampalli RY, Prevost D. Bio-encapsulation of Microbial Cells for Targeted Agricultural Delivery. Critical Reviews In Biotechnology 2011;31 211-226.

[94] Shahidi F, Han XQ. Encapsulation of Food Ingredients. Critical Reviews in Food Science and Nutrition 1993;33 501-547.

[95] Clausse D, Gomez F, Dalmazzone C, Noik C. A Method for the Characterization of Emulsions, Thermogranulometry: Application to Water-in-Crude Oil Emulsion. Journal of Colloid and Interface Science 2005;287 694-703.

[96] Iwata M, McGinity JW. Preparation of Multi-Phase Microspheres of Poly(D,L-Lactic Acid) and Poly(D,L-Lactic-Co-Glycolic Acid) Containing a W/O Emulsion by a Multiple Emulsion Solvent Evaporation Technique. Journal of Microencapsulation 1992;9 201-214.

[97] Iqbal J, Petersen S, Ulrich J. Emulsion Solidification: Influence of the Droplet Size of the Water-In-Oil Emulsion on the Generated Particle Size. Chemical Engineering & Technology 2008;34 530–534.

[98] Sajjadi, S, Jahanzad, F. Nanoparticle formation by highly diffusion-controlled emulsion polymerization,. Chemical Engineering Science 2006; 61 3001-3008.

[99] Mozafari NR, Khosravi-Darani K, Borazan GG, Cui J, Pardakhty A, Yurdugul S. Encapsulation of Food Ingredients Using Nanoliposome Technology. International Journal of Food Properties 2008;11 833–844.

[100] Fessi, H, Puisieux, F, Devissaguet, JP, Ammoury, N, Benita, S. Nanocapsule formation by interfacial polymer deposition following solvent displacement. International Journal of Pharmaceutics 1989; 55(1) R1-R4.

[101] Szente L. Stable, Controlled-Release Organophosphorus Pesticides Entrapped in B-Cyclodextrin. I. Solid State Characteristics. Journal Of Thermal Analysis And Calorimetry 1998;51 957-963.

[102] Ré MI. Microencapsulation by Spray Drying. Drying Technology 1998;16 1195-1236.

[103] Kratz K, Narasimhan A, Tangirala R, Moon SC, Revanur R, Kundu S, Kim HS, Crosby AJ, Russell TP, Emrick T, Kolmakov G, Balazs AC. Probing and Repairing Damaged Surfaces with Nanoparticle-Containing Microcapsules. Nature Nanotechnology 2012;7 87-90.

[104] Furmidge CGL, Hill AC, Osgerby JM. Physicochemical Aspects of the Availability of Pesticides in Soil. II. Controlled Release of Pesticides from Granular Formulations. Journal of the Science of Food and Agriculture. 1968;19(2) 91-95.

[105] Allan GG, Neogi AN. Diffusion From Solid Polymeric Solutions. International Pest Control 1972;14 21-27.

[106] Hussain M, Oh BY. Preparation and Study of Controlled-Release Formulations of Carbon-14 Labeled Butachlor. Toxicological And Environmental Chemistry 1991;33 101-110.

[107] Qi M, Wang F, Wang H. Study on Release Dynamics of 14C-Labeled Herbicides from Controlled-Release Formulation into Water. Henong Xuebao, 1994;8 240-246.

[108] Fernandez-Perez M, Gonzalez-Pradas E, Urena-Amate MD, Wilkins RM, Lindup I. Controlled Release of Imidacloprid From a Lignin Matrix: Water Release Kinetics and Soil Mobility Study. Journal of Agricultural and Food Chemistry 1998;46 3828 – 3834.

[109] Ritger PL, Peppas NA. A Simple Equation for Description of Solute Release II. Fickian and Anomalous Release from Swellable Devices. Journal of Controlled Release, 1987;5 37-42.

[110] Ritger PL, Peppas NA. A Simple Equation for Description of Solute Release I. Fickian and Non-Fickian Release from Non-Swellable Devices in the Form of Slabs, Spheres, Cylinders or Discs. Journal of Controlled Release, 1987;5 23-36.

[111] Villafranca-Sanchez M, Gonzalez-Pradas E, Fernandez-Perez M, Martinez-Lopez F, Flores-Cespedes F, Urena-Amate MD. Controlled Release of Isoproturon from an Alginate-Bentonite Formulation: Water Release Kinetics and Soil Mobility. Pest Management Science 2000;56 749-756.

[112] Kumar J, Singh G, Walia S, Jain S, Parmar BS. Controlled Release Formulations of Imidacloprid: Water and Soil Release Kinetics. Pesticide Research Journal 2004;16 13-17.

[113] Garrido-Herrera FJ, Gonzalez-Pradas E, Fernandez-Perez M. Controlled Release of Isoproturon, Imidacloprid, and Cyromazine from Alginate-Bentonite-Activated Carbon Formulations. Journal of Agricultural and Food Chemistry 2006:54 10053–10060.

[114] Garrido-Herrera FJ, Daza-Fernandez I, Gonzalez-Pradas E, Fernandez-Perez M. Lignin-Based Formulations to Prevent Pesticides Pollution. Journal of Hazardous Materials 2009;168 220–225.

[115] D'Antone S, Solaro R, Chiellini E, Rehab A, Akelah A, Issa R. Controlled Release of Herbicides Loaded on Oligoethylenoxylated Styrene/Divinylbenzene Resins. New Polymeric Materials 1992;3 223-236.

[116] Collins RL, Doglia S, Mazak RA, Samulski ET. Controlled Release of Herbicides. Theory. Weed Science 1973;21 1-5.

[117] Silva MS, Cocenza DS, Grillo R, Silva de Melo NF, Tonello PS, Camargo de Oliveira L, Cassimiro DL, Rosa AH, Fraceto LF. Paraquat-loaded Alginate/Chitosan Nanoparticles: Preparation, Characterization and Soil Sorption Studies. Journal of Hazardous Materials 2011;190 366-374.

[118] Tsuji K. Microencapsulation of Pesticides and Their Improved Handling Safety. Journal of Microencapsulation 2001;18 137-147.

[119] Ray PC, Yu H, Fu PP. Toxicity and Environmental Risks of Nanomaterials: Challenges and Future Needs. Journal of Environmental Science and Health, Part C: Environmental Carcinogenesis and Ecotoxicology Reviews 2009;27 1-35.

Chapter 10

Use of Biotechnology in the Control of Insects-Prague

Gleberson Guillen Piccinin, Alan Augusto Donel,
Alessandro de Lucca e Braccini,
Lilian Gomes de Morais Dan, Keila Regina Hossa,
Gabriel Loli Bazo and Fernanda Brunetta Godinho

Additional information is available at the end of the chapter

1. Introduction

Productivity gains in agriculture are satisfactory with the use of genetically modified plants and the dependency of application of insecticides on crops becomes smaller over the years. The consequences of the development and marketing of corn genetically modified (GM) have been profound, and in 2011 the area planted in the United States of America (USA) with at least one GM trait corresponded to more than 88% of was over acreage.

In addition, the efficiency gains in the production chains were only possible thanks to the entrepreneurship and management of rural producers, who adopted the most modern technologies available for science. Among these, stand: the tillage, fertilization and soil correction, the techniques of integrated management of invasive plants, diseases and insect pest and the growing adoption of improved seeds with high productive capacity. It is observed that the simple hybrids corn came to dominate the market of seeds embedded in technologies and seeds are more easily adopted by producers. This is the case of transgenic seeds, which in the culture of corn were widely adopted, including the major world producers of this cereal USA and Argentina.

In crops of corn, the losses caused by pests are limiting factor to achieve high productivity. The fall armyworm (*Spodopterafrugiperda*) is the main plague of corn in Brazil culture, causing severe damage. The attack on the plant occurs since its emergence to the booting and the silking.However, the critical period is flourishing. Losses due to Caterpillar attack may reduce production in up to 34%. Survey conducted among some 1,100 farmers who produce more than 8,000 kg ha-1 showed that, among the crops sampled, 15% received from 4 to 5 applications of insecticides and 6% received 6 to 8 applications for pest control.

Other recent surveys, conducted by Embrapa maize and sorghum, have shown that in some regions the number of applications of insecticides for the control of Caterpillar-cartridge can reach 10. In addition, there is no efficient method of chemical control for at least two other important species: or the control of the corn earworm (*Helicoverpazea*) and the Maize stalk borer (*Diatraeasaccharalis*).

The insects have been one of the biggest causes of damage in food production being these losses of the order of 20 to 30% of world production in [1]. It is estimated that approximately 67,000 species of insects cause damage to plantations and tropical regions, usually the poorest in the world, those who suffer most from the high incidence of insects-Prague in [2].

The attack of any pest depends on the development of culture, as well as the intensity of the attack, which can significantly affect the performance of the same. Chemical control is the primary measure used to prevent the immediate damage that reach the level of economic damage. Many times, the insecticides do not have the desired effectiveness and have a high cost, because are usually required multiple applications.

The insecticides used in pest control as an example in corn culture are often of low selectivity, therefore can affect the population of natural enemies, favoring the proliferation of pests and even resurgence of others. Due to these factors, the search for alternatives that can minimize or even replace the conventional insecticides was intensified and, currently, the new tactics comprise a series of alternatives: resistant plants, selective insecticides, parasitoids and entomopathogenic microorganisms. Among the entomopathogenic *Bacillus thuringiensis* (Bt), notable for its wide use in the control of pest insects of the order Lepidoptera, especially in corn culture.

Currently, transgenic production is widespread in almost all agricultural regions of the planet and with the adoption of biotechnology by greater productivity with the producers reaches lower use of insecticides. An example of this is the use of gene technology (Cry) of bacteria (Bt) in control of the main pests of maize culture. The Bt gene technology diffusion aims to make the environment more sustainable, decreasing the concentration of inert products in foods using insecticides rationally. However, many challenges must still be overcome, with that biotechnology has as a fundamental role, seek new research, sustainable in modern agriculture.

2. Importance of biotechnology in the control of insects-prague

Biotechnology is defined as a set of techniques for manipulation of living beings or part thereof for economic purposes. This broad concept includes techniques that are used on a large scale in agriculture since the early 20th century, such as tissue culture, the biological fixation of nitrogen and organic pest control. But the concept includes also modern techniques of direct modification of the DNA of a plant or a living organism, in order to change precisely the characteristics of that organism in [3].

Agricultural biotechnology varieties are used as a tool for agricultural research characterized by gene transfer of agronomic interest (and, consequently, of desired characteristics) between a donor agency (which may be a plant, a bacterium, fungus, etc.) and plants, safely.

Studies dating this year in India and China show that the Bt cotton production increased 10% to 50%, respectively, and the use of insecticides has reduced in both countries by 50%. In India, the producers have increased the income up to $ 250 or more per hectare, the farmer's income increasing national $ 840 million to $ 1.7 billion last year. Chinese farmers saw similar gains with increasing yields on average $ 220 per hectare, or more than $ 800 million nationwide. It is important to emphasize the trust farmer in technology, with 9 of 10 Indian farmers reseeding the biotech cotton 100% year after year and Chinese farmers using the technology.

It is important to recognize the need for scientific research that point on the General mode of action of Bt toxins and also that their toxicity is influenced by several factors. However, it is known that, in General, the toxicity of the Bt toxin on target organisms depends on factors such as certain pH, proteases, and the receivers in [4]. On the other hand, and more specifically, the extrinsic factors can also influence and co-factors in specific efficacy of Bt toxin on target organisms resistant and/or can also have an impact on the selectivity and toxicity to non-target organisms in [5].

Therefore, the emergence of modern biotechnology marks the beginning of a new stage for agriculture and reserves a starring role to molecular genetics. The advances in the field of plant genetics have the effect of reducing the excessive reliance on agriculture mechanical and chemical innovations, which were the pillars of the green revolution. In addition to increased productivity, modern biotechnology can contribute to the reduction of production costs, better quality foods and for the development of less aggressive to the environment in [3].

3. The *Bacillus thuringiensis* (Bt)

The *Bacillus thuringiensis*, was discovered in 1901, by the occurrence of an epidemic of mortality of larvae of the silkworm in Japan. Researchers found that it was caused by a previously unknown bacterium.

The bacterium entomopathogenic (Bt) stands out on the world stage since 1938, when the first product formulated with this pathogen was released in France.

In 1911, in Germany, the Berliner managed to isolate and characterize this bacterium, baptizing it *Bacillus* (by its cylindrical shape) *thuringiensis* (named after the German region of Turíngea). In 1938, France formulations containing right-handed bacteria colonies were sold as insecticides and, in 1954, its mode of action was discovered and its use today.

Since then more than 100 products were launched on the market and currently constitute more than 90% of gross revenues with biopesticides in [6, 7].

In some studies, this bacterium was considered inefficient in controlling *S. frugiperda* in [8, 9]. However, with the advances provided by new laboratory techniques and greater interest of researchers' positive results were obtained in [10]

The Bt a soil bacteria present in various continents, Gram-positive, aerobic and family Bacillaceae, when environmental conditions become adverse can sporulate to survive these conditions in [11]. Are found in every terrestrial environments and also in dead insects, plants and debris in [12, 13, 14, 15, 16, 17, 18, 19]. The methods to isolate this pathogen are powerful and usually easy to perform in [20, 21, 22, 23]. The number of cells obtained from Bt varied between 102 and 104 colony-forming units (CFU) per gram of soil, while in plants this number varies between 0 and ufc 100 cm^{-2} in [24].

Produces sporangia containing a endospore and crystalline inclusions of proteins that are responsible for their action entomopathogenic, among which stands out the protein CRY. This crystal is composed of a protein polypeptide called endotoxin in [25]. When larval forms of insects feed on such proteins, initiates a series of reactions that culminate with the death of the same.

4. Biotechnology vs insecticide

The insects have been one of the major causes of damage to food production in [1] and, in world terms, the losses caused by pests and diseases are quite high. The same causing losses of the order of 38% in [26]. Withdrawals in Brazil indicate that pests can be liable for loss of 2.2 billion dollars for the main Brazilian crops in [27].

Control of harmful insects is done, most of the time, by agrochemicals and, on a much smaller scale, by the employment of biological insecticides. The indiscriminate use of pesticides in combating the causal agents causes, despite its efficiency, environmental problems severe, human health, reduces number of natural enemies, and provides an accelerated selection of resistant insects in [28]. In contrast, biopesticides, Bt based, used for over a century, retainers of features less impactful on the environment and less harmful to humans ever occupied a prominent place on the market for the sale of pesticides in [29].

From the Decade of 80, because of genetic advance, it became possible to develop a new pest control strategy, which consists of the genetically modified plants resistant to insects in [and with effectiveness similar to conventional insecticides in [31, 32].

The first experiments with genetically modified (GM) plants were made in 1986, in the United States and in France. The first variety marketed a vegetable species produced by genetic engineering was the "FlavrSavr Tomato ", developed by the American company Calgene and marketed from 1994.

Between 1987 and 2000 there were more than 11,000 field trials in 45 countries and cultures more frequently tested were corn, tomatoes, soybeans, canola, potatoes and cotton, and the genetic features introduced were herbicide tolerance, product quality, virus-resistance and resistance to insects in [33].

Rank	Country	Area (million hectares)	Biotech Crops
1	USA*	69.0	Maize, soybean, cotton, canola, sugarbeet, alfafa, papaya, squash
2	Brazil*	30.3	Soybean, cotton, maize
3	Argentina*	23.7	Soybean, cotton, maize
4	India*	10.6	Cotton
5	Canada*	10.4	Canola, maize, soybean, sugarbeet
6	China*	3.9	Cottton, papaya, poplar, tomato, sweet paper
7	Paraguay*	2.8	Soybean
8	Pakistan*	2.6	Cotton
9	South América*	2.3	Maize, soybean, cotton
10	Uruguay*	1.3	Soybean, maize
11	Bolivia*	0.9	Soybean
12	Australia*	0.7	Canola, cotton
13	Philippines*	0.6	Maize
14	Myanmar*	0.3	Cotton
15	Burkina Faso*	0.3	Cotton
16	Mexico*	0.2	Cotton, soybean
17	Spain*	0.1	Maize
18	Colombia	<0.1	Cotton
19	Chile	<0.1	Maize, soybean, canola
20	Honduras	<0.1	Maize
21	Portugal	<0.1	Maize
22	Czech Republic	<0.1	Maize
23	Poland	<0.1	Maize
24	Egypt	<0.1	Maize
25	Slovakia	<0.1	Maize
26	Romania	<0.1	Maize
27	Sweden	<0.1	Potato
28	Costa Rica	<0.1	Cotton, soybean
29	Germany	<0.1	Potato
Total		160.0	

* 17 biotech mega-countries growing 50,000 hectares, or more, of biotech crops.

** Rounded off to the nearest hundred thousand

Source: in [34].

Table 1. Global Area of Biotech Crops in 2011: by Country (Million Crops)**.

These days, according to the annual report of 2011 on the use of transgenic crops, the non-profit organization International Service for the acquisition of AgriBiotech Applications (ISAAA) observed an increase of 94 times in planted area of 1.7 million hectares in 1996 to 160 million hectares in 2011 (Table 1), allowing biotech crops become more agricultural technology adopted in the history of modern agriculture.

The endless search for alternative methods of insect control-Prague has been held strongly by several research groups worldwide, due to the need of a more sustainable agriculture and more committed to environmental preservation in [35].

In this way, farmers have adopted this technology Bt targeting an increasing effective production to sustainable agriculture in [36]. The benefits of this technology are: reduction of environmental effects on toxins, safety in use, efficiency, conservation of natural enemies and reduction of fungal diseases.

The first advantage is the production of protein *Cry*, by plants-Bt, which is not affected by environmental factors such as atmospheric fallout, light incidence, and high temperatures in [37]. In addition, the homogeneity of the protein, in plant tissues, allows a more efficient use of insecticide effect than the application (spraying) of biopesticides, Bt, based on plants. The second advantage is the possibility of a higher level of security in relation to insecticide formulated because the proteins and does not accumulate in fatty tissues, are not toxic to humans and pets. Tied to these characteristics, the protein Cry, has no activity by contact, being necessary, the ingestion of the toxin by the insect, to have the effect of insecticide. The third advantage is the *Heliothisvirescens*control significant and *Pectinophoragossypiella*, for example, in Bt cotton culture, between 95 and 99% efficiency in [38]. The fourth advantage is the preservation of natural enemies, therefore, secondary pests can become a problem if the population of beneficial insects is reduced by the use of chemical insecticides of low selectivity. The fifth benefit, no less important, is the reduction of fungal diseases. The lesions caused by insects, in the organs of plants, fungi infection, create opportunities mainly in the genus *Fusarium e Aspergillus*in [39]. The primary importance of these fungi is the presence of micotoxins, particularly fumosinsandaflatoxins produced by them. The fumosins can be fatal to horses and pigs in [40]. And aflatoxinis extremely toxic to animals and humans in [39]. The dramatic reduction of insect attack, leads to reduction of insect attack, and consequently, decreases the production of micotoxins.

5. Mode of action of protein Cry

Currently the insect resistant transgenic plants expressing genes, inductors, an insecticidal protein called Cry, derived from the bacterium *Bacillus thuringiensis*(Bt). The mechanisms by which proteins exert their effect are *Cry* elucidated by pore formation model discussed below:

The mode of action of Cry proteins, produced by the plant, it is accomplished, orally, by susceptible insect. The process begins by solubilization of crystals in alkaline pH around 9.5, in the gut of insects, releasing protoxin of 130 kDa to Cry1 and Cry2 to 79kDa.. After this breakdown, the protoxinare activated by digestive enzymes, forming toxic fragments of 60-65 kDa. These monomers bind to receptors specific primary, located in the apical membrane of the microvillus membranes of the columnar cells of the intestine of the larva. It is in this step that the affinity between the toxin and the receiver, for example, Cry1ae protein, lepdopteros is recognized as an important factor in determining the spectrum insecticidal Cry proteins. Later, the monomers bind to secondary receivers, which are proteins ancoradorasglicosil-phosphatidyl-inisitol (API), as phosphates and alkaline, to the lepidoptero*Heliothisvirences*. After this binding, the now oligômerose inserts into the membrane, where there are receptors for API, and leads to the formation of pores in the cell membrane of the intestinal epithelium and therefore destruction of microvilli membranes, hypertrophy of epithelial cells, vacuolization of cytoplasm, cell lyses and intestinal paralysis/death of the insect in [11, 41].

6. The safety of the use of Btplants

In relation to the safety of the use of Bt corn plants as an example, several tests are conducted to certify the safety of its use in the environment and in food and feed. Initially, the protein *Cry* is tested in animal models, such as rats and mice, for the verification of the toxic potential. One of these tests, called acute toxicity consists in forced ingestion of a pure animal protein by solution and on the observation of effects of this. The product only goes to the next steps of assessment if no effect is observed and diagnosed.

We can cite as an example the test performed with Safety, this protein in maize Herculex®. This protein was tested on mice to the level of 576 mg per kilogram of live weight and no side effect was observed. To be exhibited at similar level, a person weighing 70 kg would take almost 5 tons of raw corn grains. This without taking into consideration the aspect that the human digestive tract, not to have alkaline pH, would not be able to downgrade this protein crystal.

Other reviews include the potential to cause allergies as well as the corn grain consumption by other animals such as chicken and fish, and what is called substantial equivalence, which is comparing the nutritional profile of the genetically modified maize with conventional maize. The corn will only be released commercially and, therefore, will go to the market when, in these analyses, the nutritional content between the conventional and transgenic corn were exactly the same, except, of course, the presence of protein inserted.

In the analysis of environmental safety, non-target organisms, how insects from another order, class or species, natural enemies and beneficial insects like bees, for example, are exposed to proteins inserted or the pollen grains that express and are evaluated its effects.

If all tests present results within expected ranges and be proven that there is no risk of harm or damage to health and the environment, these damages are compiled and submitted to the competent authorities of the country where you intend to market the product for analysis and approval of use and consumption. In Brazil these analyses are made by the national technical Commission on Biosafety (CTNBio) and the approval of a product in one country does not guarantee that the same is approved in another. For example, the event MON810 (Yield Gard®) was approved in 1996 in the United States, in 1998 in Argentina and only in 2007 in Brazil.

For the use of Bt corn, just the producer, in addition to using the seeds of biotech corn, fulfill two rules: the coexistence, required by law, and the rule of Insect Resistance management (MRI), recommended by (CTNBio).

The coexistence rule requires the use of a 100 m isolating surround of transgenic maize plantations of corn to retain without transgenic contamination. Alternatively, you can use a surround of 20 m, provided they are sown maize transgenic not 10 ranks (equal-sized and transgenic maize cycle) isolating the area of transgenic maize.

The CTNBiorecommendation for Insect Resistance management is the use of the area of refuge. This recommendation is the result of consensus that the cultivation of Bt corn in large areas will result in the selection of biotypes of target pests resistant to Bt toxins. Obviously, the monitoring of the infestation of plants is also important because, depending on the used hybrid and intensity of infestation, the producer may need to adopt additional control measures.

The biggest concern with the use of Bt corn is on transgenic crops and coexistence of transgenic crops do not. Coexistence is the set of agricultural practices allowing farmers grain production from conventional transgenic and organic crops, according to standards of purity and to meet legal requirements for labeling. The adoption of the rules of coexistence is essential to preserve the freedom of choice of producers and consumers. Coexistence is also a topic particularly relevant when there is market incentive for the provision of non-transgenic maize. Evidence of their practical viability is the coexistence of a considerable number of different varieties of open pollination still in use.

Showed that companies in possession of this technology must guide growers on the rules of coexistence. The producer also held technical information, stick to them properly and conscious.

Information on packages of seed of Bt corn, there is a contract in which the producer, to open it, assumes the responsibility of following the rules of coexistence and the resistance management. Therefore, it is incumbent upon the producer responsibility of use of these rules. It is important to remember that the incorrect use of technology can take it to ineffectiveness in little time. If the producer is interested in paying more for Bt corn seed, is because he believes in the benefits that this technology is bringing to your production system. Therefore, it must be motivated to use this technology in a responsible way (using the area of refuge), to take ownership of this benefit for much longer.

In relation to Bt cotton to China is the leader in this technology. In 2006 6.3 million farmers, or more than 60% of the number of farmers who have sown transgenic in the world in [42].

China is one of the only exceptions in the world to require shelters, although this may be changing. Even though the refuges were a way to reduce the accumulation of resistance to Bt toxin, the large number of small properties makes this strategy is difficult to apply.

The use of refuge in a developing country such as China becomes a challenging activity. Studies on policies of refuge on a large scale, in extensive agricultural systems of the United States of America, show that monitoring and implementation costs are negligible. Although this practice is reasonable in extensive production systems and with a small number of farms, they may not be suitable in developing countries. In developing countries like China, the agriculture sector is fragmented into millions of smallholdings, where each family has a diverse set of cultures in [43].

As a result, it is likely that the implementation of the strategy of refuge to the style (IE, all farmers planting Bt cotton are forced to grow cotton non-Bt with refuge) would require a large implementation effort, making these types of strategies of refuge becomes unviable unless farmers received individual incentives to implement refuges based on self-interest. This is unlikely, since the build-up of resistance to Bt technology is a collective evil (compared to the more common public as well) that is unlikely to be accounted for by individuals in [43].

The area of refuge is the sowing of 10% of the area planted with Bt corn hybrids using Btnot equal size and cycle, preferably their isogenic hosts. The area of refuge should not be more than 800 m away from the transgenic plants. This is the average distance by dispersion of adults of LCM in the field in [44].

All these recommendations are in order to synchronize the intersections of potential adults surviving in area of Bt corn with which emergency in the area of refuge. The structured refuge must be drawn according to the acreage with the Bt corn to plots dimensions above 800 m in the shortest side (or Ray), cultured with Bt corn refuge will be needed in their tracks internal plots. Yet, according to the recommendation, in the area of refuge CTNBiois allowed the use of other methods of control, provided that they are not used Btbased bioinsecticide.

7. Future trends for the Bt technology

The worst drought in more than half a century in corn-producing region of the United States should reduce the crop in that country at the lowest level in five years, where their stocks will be reduced to the lowest level in 17 years. The initial productivity is bad in the few fields harvested in areas of the Midwest, which represents 75 percent of the area with corn and soybeans in the United States. With this the world, returned his eyes to the Brazilian corn crop this year had one of the largest capacities of the whole story.

Brazilian agriculture won in the early 1980, an important milestone and helped the country to assume the rank among the major food producers in the world. Called when the off-season summer pós-safra was used by the producers for the planting of corn seeds uncultivated, and subsequently became part of the farmer's strategy to increase your productivity.

In more than 30 years of history, the off-season if expanded, gained strength and hit record. According to the 9th Brazilian harvest survey of grain (2011/2012), released by the national supply company (Conab) in June, the area planted with corn in the off-season is estimated at 7.188 million hectares, number 22% higher than last year's off-season.

According to Conab, the number is explained by good price prospects for climate advantage provided with the anticipation of rains for planting, and by the good harvest of soybeans, which encouraged producers to extend their crops.

Most producers that have soy as flagship summer crop production, bet on corn cultivation off-season, with attractive price and the advent of biotechnology has been the increase of productivity and safety in pest management and, in addition, there were the intangible gain with the decrease of insecticide applied in the environment.

The good news for the cultivation of corn in Brazil, according to Conab, are linked to the main corn producing States off-season: MatoGrosso, Paraná and MatoGrosso do Sul, which added to the total cultivated in the past year, the areas of 732.7 thousand, 283.4 billion and 193.2 thousand hectares, respectively.

With the data of the survey, the company foresees a production of about 32.9 million tones for corn second crop, or 53.1% to 21.5 million achieved last season. The Brazilian farmer realizes, each year, the off-season is a good deal, that is, it is an opportunity to increase the profitability of farming, maximizing the use of resources already invested.

Logically that this increased production, requires a quick response companies to address the needs of new hybrids of corn, which led to the increase of releases in this area, where the market turned to the specific needs of each region. With this, the producers have the opportunity to plant the best genetic, associated with the best biotechnology.

According to the Brazilian Association of producers of corn (Abramilho), the winter harvest has been growing a lot for two reasons: first, because soy has open space for the cultivation of this crop and, second, because of the technology. The conventional corn planting and with few seeds has become the past, and today, the use of increasingly technology for these cycles has achieved nearly the same results of the summer harvest.

Between farmers, the assessment is that the off-season will consolidate its position as an important complement in income and must, year after year, to expand to areas that do not yet have this established planting. There is a very strong demand for producer hybrids with more technology. Therefore, in recent years, companies have expanded investment in research in Brazil to bring to market the best product for the features of each region.

With the events of biotechnology of the culture of corn producers expect to achieve greater productivity per area, plants tolerant to various events and reduce production costs, primarily related to less use of pesticides.

In the case of Bt crops (which produce a toxin in their cells), their adoption allowed a reduction of 56 million kg of insecticides between 1996 and 2011. In General, the transgenic seed calculation led to decrease of 183 million tonnes in the use of pesticides. In Brazil, the Bt seed companies are also newly tolerant to herbicides, thereby opening a new path of efficiency in the management of pests and weeds, factors that interfere with productivity and steal the producer's profit.

Due to various factors the area planted with genetically modified seeds should reach 36, 6milhões hectares in the next harvest, second 1° monitoring of adoption of agricultural crop 2012/13. The forecast points to a 12.3% higher adoption in comparison to the previous year and means 4 million new acres with transgenic varieties.

The leadership in adopting biotechnology continues with soybeans, which must have 88.1% of crops with genetically Modified seeds, an area estimated at 23.9 million hectares. And corn, which begins to cultivate the fourth crop with transgenic hybrids, already approaching that level. The winter crop represents the second highest rate of adoption, with 87.8%, or 6.9 million hectares of transgenic seeds. In the case of the summer harvest, the adoption must represent 62.6% of the total area or 5, 2 million hectares.

The cotton must have 50.1%, or 546 thousand hectares of the total area with transgenic seeds. The continuous growth of adoption of biotechnology should be attributed to the increase of new varieties available in the market and that, today, are adapted to the different agricultural areas of the country. The direct and indirect benefits arising from the use of these seeds have been singled out by farmers as one of the biggest reasons for choice. In relation to States, MatoGrosso follows in the lead, with 9, 6milhões hectares, followed by Paraná with 6.6 million hectares. Herbicide tolerance technology follows in the lead with 25, 3milhões hectares, followed by seeds with resistance to insects, with 5.7 million hectares, and the gene technology, combined with 5.6 million hectares.

All these good news coming from the field are a major impasse regarding the prices of maize, companies that buy corn in Brazil must face an even more difficult year in terms of price in 2013, compared with 2012, and projected smaller cereal availability in 2012/13.

In the first half of 2012 prices were behaved and there was even falling prices to the extent that it was becoming clear that we would have a great off-season.

Cereal prices in the international market began to rise and reached record levels in recent weeks on the Chicago Stock Exchange in function from the perspective of large us crop failure caused by the worst drought in more than 50 years. Prices in Brazil are now strongly tied to international prices due to a large demand from international buyers. Certainly we have a Brazilian corn buyer pressure by international customers, as strong or stronger than this year. Because the major supplier of the world's corn, which are the United States, will have less exportable surplus in history.

Corn futures in Chicago (CBOT) reached the highest value of all time before disclosure of the report of the United States Department of agriculture (USDA), which should cut forecasts for USA crops this year.

The contract of the new crop, basis December reached the peak of 8.2975 dollars a bushel, the highest ever recorded in the Chicago Stock Exchange and above the previous record of 8.2875 dollars per busheltested three weeks ago by September.

As we saw in this chapter to biotechnology is a very important tool for the development of Brazil as world agricultural power, but we must emphasize the importance of research. And to get an idea of its importance, the increase in productivity in the various cultures saved 60 million hectares to Brazil, but still in some cultures our average productivity is low. In corn, for example, our productivity is half of U.S.A productivity. Search is a technology and factor income generator to the field, but Brazil has employed little recourse in the area. In 2011 employed 1.3% of GDP in science and technology, and in 2012 must employ only 0.9%, which is a setback and lack of objectivity.

8. Perspectives and final considerations

The biggest challenge of this century is to feed the whole world's population; poverty and hunger are inextricably linked and are about 1 billion people, mainly rentals. So what we will do to overcome this in a sustainable way? What are the threats to the production, distribution and safety of these foods? And, mainly, what we will do to allow people to have access to them?

In 2011, approximately half of the world's poor were small resource-poor farmers, whilst another 20% were the rural landless who are completely dependent on agriculture for their livelihoods. Thus, 70% of the world's poor are dependent on agriculture – some view this as a problem, however it should be viewed as an opportunity, given the enormous potential of both conventional and the new biotechnology applications to make a significant contribution to the alleviation of poverty and hunger and to doubling food, feed and fiber production by 2050.

In the next fifty years will be nine billion people and the world will consume twice as much food as the world has consumed since the beginning of agriculture 10.000 years ago. The challenge of feeding the world and interest to increase the potential of biotechnology are intimacies trailers and, as biotech crops already occupy about 160 million hectares or 10% of the world's arable land, is significant and visible the ancestry of this market in today's society.

According to some international institutes, the world until 2020 will grow 20% in food production and Brazil will have a 40% growth in production with an increase of just 16% of the area. To see the responsibility, since the increment of production in traditional countries like the USA will be 15%, China 15%, and in the whole of Europe, 4%.

With all this challenge, it should be noted that the transgenic plants are not panacea to solve all the problems of agriculture. Transgenic agriculture is only a complement to conventional agriculture, organic and other modalities. However the endless search for improvement of this tool will provide future generations, guarantees of more sustainable living conditions and higher quality food.

Author details

Gleberson Guillen Piccinin, Alan Augusto Donel, Alessandro de Lucca e Braccini, Lilian Gomes de Morais Dan, Keila Regina Hossa, Gabriel Loli Bazo and Fernanda Brunetta Godinho

Department of Agronomy, State University of Maringá, Maringá, Paraná, Brazil

References

[1] Estruch, J.J., et al. Transgenic plants: an emerging approach to pest control. Nature Biotechnology, New York, v.15, p.137-141, 1997.

[2] Herrera-Estrella, L. Transgenic plants for tropical regions: Some consideration about their development and their transfer to the small farmer. Proceedings of the National Academy of Sciences of the USA, Washington, 96:5978-5981, 1999.

[3] Silveira, J. M. F. J.; Borges, I. C.; Buainain, A. M. Biotecnologia e agricultura: da ciência e tecnologia aos impactos da inovação. São Paulo Perspectiva. v.19, n. 2, p. 101-114. 2005.

[4] Maagd, R. A.; Bravo, A. How Bacillus thuringiensis has evolved specific toxins to colonize the insectworld. Trends Genet 17:193–199, 2001.

[5] Sharma, H. C.; Sharma, J. H.; Crouch, K. K.; Genetic transformation of crops for insect resistance: potential and limitations. Critical Reviews in Plant Sciences. 23:47–72, 2004.

[6] Beegle, C. B.; Yamamoto, T. Invitation paper (C.P. Alexander Fund): History of Bacillus thuringiensisBerliner research and development. Canadian Entomologist, Toronto, v.124, p.587-616, 1992.

[7] Glare, T. R.; O'callaghan, M. Bacillus thuringiensis: biology, ecology and safety. Chichester: John Wiley & Sons, 2000. 350 p.

[8] Lima, J. O. G.; Zanuncio, J. C. Controle da "lagarta do cartucho do milho", Spodoptera frugiperda, pelo carbaril, carbofuram, dipel (Bacillus thuringiensis) e endossulfam. Revista Ceres, Viçosa, v.23, n.127, p.222-225, 1976.

[9] Garcia, M. A.; Simões, M.; Habib, M. E. M. Possible reasons of resistance in larvae of Spodopterafrugiperda (Abbot & Smith, 1797) infected by Bacillus thuringiensisvar. kurstaki. Revista de Agricultura, Piracicaba, v. 57, p. 215-222,1982.

[10] Hernandez, J. L. L. Évaluation de la toxicité de Bacillus thuringiensis surSpodoptera frugiperda. Entomophaga, Paris, v.32, p.163-171, 1988.

[11] Schnepf, E.; Crickmore, N.; Vanrie, J.; Lereclus, D.; Baum, J.; Fietelson, J.; Ziegler, D. R.; Dean, D. H. Bacillus thuringiensis and its pesticide crystal proteins. Microbiology Molecular Biology Reviews, 62, 775-806,1998.

[12] Bernhard, K.; Jarrett, P.; Meadows, M. Natural isolates of Bacillus thuringiensis: worldwide distribution, characterization, and activity against insect pests. Journal ofInvertebrate Pathology, Orlando, 70, 59-68, 1997.

[13] Hossain, M. A.; Ahmed, S.; HoqueS. Abundance and distribution of Bacillus thuringiensisin the agricultural soil of Bangladesh. Journal of Invertebrate Pathology, Orlando, v.70, p. 221-225, 1997.

[14] Forsyth, G.; Logan, N. A. Isolation of Bacillus thuringiensisNorthern Victoria Land, Antarctica. Letters in Applied Microbiology, v. 30, p. 263-266, 2000.

[15] Hongyu, Z.; Ziniu, Y.; Wangxi, D. Isolation, distribution and toxicity of Bacillus thuringiensisfrom warehouses in China. Crop Protection, Amsterdam, v.19, p.449-454, 2000.

[16] Maeda, M.; Mizuki, E.; Nakamura, Y. Recovery of Bacillus thuringiensisfrom marine sediments of Japan. Current Microbiology, Washington, v.40, n.6, p.418-422, 2000.

[17] Martínez, C.; Caballero, P. Contents of cry genes and insecticidal activity of Bacillus thuringiensisstrains from terrestrial and aquatic habitats. Journal of Applied Microbiology, 92:745-752, 2002.

[18] Uribe, D.; Marinez, W.; Cerón, J. Distribution and diversity of cry genes in native strains of Bacillus thuringiensisobtained from different ecosystems from Colombia. Journal of Invertebrate Pathology, v.82, p.119-127, 2003.

[19] Wang, J.; Boets, A.; Van Rie, J. et al. Characterization of cry1, cry2 and cry9 genes in Bacillus thuringiensis isolates from China. Journal of Invertebrate Pathology, v.82, p. 63-71, 2003.

[20] Saleh, S. M.; Harris, R. F.; Allen, O. N. Method for determining Bacillus thuringiensisvar. thuringiensis Berliner in soil. Canadian Journal of Microbiology, v.15, p. 1101-1104, 1969.

[21] World Health Organization, 1985. Informal Consultation on the Development of Bacillus sphaericus as a microbial larvicide. Genebra, UNDP/World Bank/ WhoSpecialProgramme for Research and Training in Tropical Deseases.

[22] Travers, R. S., Martin, P. A. W.; Reichefelder, C. F. Selective process for efficient isolation soil Bacillus sp. Applied and Environmental Microbiology, v.53, p.1263- 1266, 1987.

[23] Swiecicka, I.; Fiedoruk, K.; Bednarz, G. The occurrence and properties of Bacillus thuringiensis isolated from free-living animals. Letters in Applied Microbiology, v. 34, p. 194-198, 2002.

[24] Damgaard, P. H. Natural occurrence and dispersal of Bacillus thuringiensisin the environment. In: Charles, J.F.; Delécluse, A.; Nielsen-Le Roux, C. Entomopathogenic bacteria: from laboratory to field application. Netherlands: Kluwer Academic Publishers, 2000. p. 23-40.

[25] Navon, A. Control of lepidopteran pests with Bacillus thuringiensis. In: ENTWISTLE, P. F.; CORY, J. S.; BAILEY, M. J.; HIGGS, S. Bacillus thuringiensis, an environmental biopesticide: theory and practice. Chichester: Wiley, 1993. 311p.

[26] Gallo, D.; Nakano, O.; Silveira, S.; Carvalho, R. P. L.; Batista, G. C.; Berti Filho, E.; Parra, J. R. P.; Zucchi, R. A.; Alves, S. B.; Vendramin, J. D.; Marchini, L. C.; Lopes, J. R. S.; Omoto, C. Entomologia agrícola. Piracicaba: Fealq, 2002. 920 p.

[27] Bento, J. M. S. Perdas por insetos na agricultura. Ação Ambiental II, v.4, p. 19-21,1999.

[28] Arantes, O. M. N.; Vilas-Bôas, L. A.; Vilas-Bôas, G. T. Bacillus thuringiensis: Estratégias no controle Biológico. In: Serafinil. A.; Barros, N. M.; Azevedo, J. L. (Eds.). Biotecnologia: Avanços na agricultura e agroindústria. Caxias do Sul: Educs. v. 2, p. 269-293, 2002.

[29] Navon, A. Bacillus thuringiensis insecticides in crop protection. reality and prospects. Crop Protection, Oxford, v.19, p.669-676, 2000.

[30] Armstrong, C. L.; Parker, G. B.; Pershing, J. C.; Brown, S. M.; Anders, P. R.; Duncan, D. R.; Stone, T.; Dean, D. A.; Deboer, D. L.; Hart, J.;Ooe, A. R.;Morrish, F. M.; Pajau, M. E.; Petersen,W. L.; Reich, B. J.; Rodriguez, R.;Santino, C. G.; Sato, S. J.; Schuler, W.; Sims, S. R.; Stehling, S.; Tarochione, L.J.; Fromm, M. E. Field evaluation of European corn borer control in progeny of 173 transgenic corn events expressing an insecticidal protein from Bacillus thuringiensis. Crop Science, Madison, v. 35, p. 550-557. 1995.

[31] Fischhoff, D. A. Insect-resistant Crop Plants. In: Persley, G.J. (Ed.). Biotechnology and integrated pest management. Wallingford: CAB INTERNATIONAL, 1996. cap.12, p. 214-227.

[32] Paoletti, M.G.; Pimentel, D. Environmental risks of pesticides versus genetic engineering for agricultural pest control. Journal of Agricultural and Environmental Ethics, v.12, p.279-303, 2000.

[33] Borém, A.; Santos, F. R. Biotecnologia Simplificada. Viçosa: Ed. UFV, 2001.

[34] James, C. 2011. Global Status of Commercialized Biotech/GM Crops: 2011. ISAAA Brief No. 43. ISAAA: Ithaca, NY. Para informações ou solicitações de outras publicações da ISAAA sobre a agricultura biotecnológica em países em desenvolvimento, 2011.

[35] Bobrowski, V.L; Fiuza, L. M.; Pasquali, G.; Bodanesezanettini, M. H. Genes de Bacillus thuringiensis: uma estratégia para conferir resistência a insetos em plantas. Ciência Rural; 34 (1): 843-850, 2002.

[36] Dunwell, J. M. Transgenic crops: the next generation, or an example of 2020 vision. Annals of botany, kent, v.84, p.269- 277, 999.

[37] Betz, F. S.; Hammond, B. G.; Fuchs, R. L. Safety and advantages of Bacillus thuringiensis-protected plants to control insect pests. Regulatory, Toxicology and Pharmacology, San Diego, v. 32, p.156-173, 2000.

[38] Halcomb, J. L. et al. Survival and growth of bollworm and tobacco budworm nontransgenic and transgenic cotton expressing CryIAc insecticidal protein (Lepidoptera :Noctuidae). Environmental Entomology, Lanham, v. 25, n. 2, p.250-255, 1996.

[39] Munkvold, G. P.; Hellmich, R. L. Comparison of fumosin concentrations in kernel of transgenic Bt maize hybrids and nontransgenic hybrids. Plant Disease, St. Paul, v.83, p.130-138,1999.

[40] Marasas, W. F. O. Fumosins: their implications for human andanimal health. Natural Toxins, Somerset, v.3, p.193-198, 1995.

[41] Bravo, A.; Gomez, I.; Conde, J.; Muçoz-Garay,C.; Sanchez,J.; Miranda, R; Zhuang,M.; Gill, S. S.; Sobern'n, M. Oligomerization triggers binding of a Bacillus thuringiensisCry1Ab pore-forming toxin to aminopeptidase N receptor leading to insertion into membrane microdomains. BiochimicaetBiophysicaActan, p. 38-46, 2004.

[42] James, C. (2006). Global Status of Commercialized Transgenic Crops: 2006. The International Services Acquisition of Agri-Biotech Applications (ISAAA), Brief, n.35: Preview. Ithaca, NY: ISAAA.

[43] Qiaoa, F.; Wilenb, J.; Huangc, J.; Rozelle, S. Dynamically optimal strategies for managing the joint resistance of pests to Bt toxin and conventional pesticides in a developing country. European Review of Agricultural Economics. v. 36, p. 253–279, 2009.

[44] Waquil, J. M.; Vilella, F. M. F.; Foster, J. E. Resistência do milho (Zea mays L.) transgênico (Bt.) à lagarta-do-cartucho, Spodoptera frugiperda (Smith) (Lepidóptera: Noctuidae). Revista Brasileira de Milho e Sorgo, Sete Lagoas, v. 1, n. 3, p. 1-11, 2002.

Permissions

The contributors of this book come from diverse backgrounds, making this book a truly international effort. This book will bring forth new frontiers with its revolutionizing research information and detailed analysis of the nascent developments around the world.

We would like to thank Stanislav Trdan, for lending his expertise to make the book truly unique. He has played a crucial role in the development of this book. Without his invaluable contribution this book wouldn't have been possible. He has made vital efforts to compile up to date information on the varied aspects of this subject to make this book a valuable addition to the collection of many professionals and students.

This book was conceptualized with the vision of imparting up-to-date information and advanced data in this field. To ensure the same, a matchless editorial board was set up. Every individual on the board went through rigorous rounds of assessment to prove their worth. After which they invested a large part of their time researching and compiling the most relevant data for our readers. Conferences and sessions were held from time to time between the editorial board and the contributing authors to present the data in the most comprehensible form. The editorial team has worked tirelessly to provide valuable and valid information to help people across the globe.

Every chapter published in this book has been scrutinized by our experts. Their significance has been extensively debated. The topics covered herein carry significant findings which will fuel the growth of the discipline. They may even be implemented as practical applications or may be referred to as a beginning point for another development. Chapters in this book were first published by InTech; hereby published with permission under the Creative Commons Attribution License or equivalent.

The editorial board has been involved in producing this book since its inception. They have spent rigorous hours researching and exploring the diverse topics which have resulted in the successful publishing of this book. They have passed on their knowledge of decades through this book. To expedite this challenging task, the publisher supported the team at every step. A small team of assistant editors was also appointed to further simplify the editing procedure and attain best results for the readers.

Our editorial team has been hand-picked from every corner of the world. Their multi-ethnicity adds dynamic inputs to the discussions which result in innovative outcomes. These outcomes are then further discussed with the researchers and contributors who give their valuable feedback and opinion regarding the same. The feedback is then collaborated with the researches and they are edited in a comprehensive manner to aid the understanding of the subject.

Apart from the editorial board, the designing team has also invested a significant amount of their time in understanding the subject and creating the most relevant covers. They scrutinized every image to scout for the most suitable representation of the subject and create an appropriate cover for the book.

The publishing team has been involved in this book since its early stages. They were actively engaged in every process, be it collecting the data, connecting with the contributors or procuring relevant information. The team has been an ardent support to the editorial, designing and production team. Their endless efforts to recruit the best for this project, has resulted in the accomplishment of this book. They are a veteran in the field of academics and their pool of knowledge is as vast as their experience in printing. Their expertise and guidance has proved useful at every step. Their uncompromising quality standards have made this book an exceptional effort. Their encouragement from time to time has been an inspiration for everyone.

The publisher and the editorial board hope that this book will prove to be a valuable piece of knowledge for researchers, students, practitioners and scholars across the globe.

List of Contributors

Mauro Prato
Dipartimento di Genetica, Biologia e Biochimica, Facolta' di Medicina e Chirurgia, Universita'di Torino, Italy
Dipartimento di Neuroscienze, Facolta' di Medicina e Chirurgia, Universita' di Torino, Italy

Manuela Polimeni
Dipartimento di Genetica, Biologia e Biochimica, Facolta' di Medicina e Chirurgia, Universita'di Torino, Italy

Giuliana Giribaldi
Dipartimento di Genetica, Biologia e Biochimica, Facolta' di Medicina e Chirurgia, Universita'di Torino, Italy

María-Lourdes Aldana-Madrid
Departamento de Investigación y Posgrado en Alimentos, Universidad de Sonora. Hermosillo, Sonora, México

María-Isabel Silveira-Gramont
Departamento de Investigación y Posgrado en Alimentos, Universidad de Sonora. Hermosillo, Sonora, México

Guillermo Rodríguez-Olibarría
Departamento de Investigación y Posgrado en Alimentos, Universidad de Sonora. Hermosillo, Sonora, México

Fabiola-Gabriela Zuno-Floriano
Department of Environmental Toxicology, University of California, Davis, California, USA

Andréia da Silva Almeida
Seed Care Institute- Syngenta, Brazil

Francisco Amaral Villela
PPG Ciência e Tecnologia de Sementes, Universidade Federal de Pelotas, Brazil

João Carlos Nunes
Syngenta Crop Protection- Seed Care Institute, Brazil

Geri Eduardo Meneghello
Universidade Federal de Pelotas, Brazil

Adilson Jauer
Syngenta Crop Protection, Brazil

T.S. Imo
Department of Chemistry, Graduate School of Engineering and Science, University of the Ryukyus, Nishihara, Okinawa, Japan
Faculty of Science, National University of Samoa, Samoa

T. Oomori
Department of Chemistry, Graduate School of Engineering and Science, University of the Ryukyus, Nishihara, Okinawa, Japan

M.A Sheikh
Department of Chemistry, Graduate School of Engineering and Science, University of the Ryukyus, Nishihara, Okinawa, Japan
Research Unit, the State University of Zanzibar, Tanzania

T. Miyagi
Okinawa Prefectural Institute of Health and Environment, Ozato Ozato Nanjo, Okinawa, Japan
Water Quality Control Office, Okinawa Prefectural Bureau, Miyagi, Chatan, Okinawa, Japan

F. Tamaki
Water Quality Control Office, Okinawa Prefectural Bureau, Miyagi, Chatan, Okinawa, Japan

Francisco Sánchez-Bayo
University of Technology Sydney, Australia

Henk A. Tennekes
Experimental Toxicology Services (ETS) Nederland BV, The Netherlands

Koichi Goka
National Institute for Environmental Sciences, Japan

Bennett W. Jordan, Barbara E. Bayer, Philip G. Koehler and Roberto M. Pereira
Department of Entomology and Nematology, Steinmetz Hall, Natural Area Drive, University of Florida, Gainesville, FL, USA

Papadopoulos Elias
Laboratory of Parasitology and Parasitic Diseases, Faculty of Veterinary Medicine, Aristotle University of Thessaloniki, Greece

Christopher J. Fettig
Pacific Southwest Research Station, USDA Forest Service, Davis, CA, USA

Donald M. Grosman
Forest Health, Texas A&M Forest Service, Lufkin, USA

A. Steven Munson
Forest Health Protection, USDA Forest Service, Ogden, UT, USA

Bruno Perlatti, Patrícia Luísa de Souza Bergo, Maria Fátima das Graças Fernandes da Silva, João Batista Fernandes and Moacir Rossi Forim
Federal University of São Carlos, Department of Chemistry, Brazil

Gleberson Guillen Piccinin, Alan Augusto Donel, Alessandro de Lucca e Braccini, Lilian Gomes de Morais Dan, Keila Regina Hossa, Gabriel Loli Bazo and Fernanda Brunetta Godinho
Department of Agronomy, State University of Maringá, Maringá, Paraná, Brazil

Printed in the USA
CPSIA information can be obtained
at www.ICGtesting.com
JSHW011430221024
72173JS00004B/743